T0340229

Statistical Design of Experiments with Engineering Applications

STATISTICS: Textbooks and Monographs

Recent Titles

Statistical Design of Experiments with Engineering Applications

Kamel Rekab and Muzaffar Shaikh
Florida Institute of Technology

CRC Press
Taylor & Francis Group
Boca Raton London New York

CRC Press is an imprint of the
Taylor & Francis Group, an **informa** business
A CHAPMAN & HALL BOOK

CRC Press
Taylor & Francis Group
6000 Broken Sound Parkway NW, Suite 300
Boca Raton, FL 33487-2742

First issued in paperback 2019

© 2005 by Taylor & Francis Group, LLC
CRC Press is an imprint of Taylor & Francis Group, an Informa business

No claim to original U.S. Government works

ISBN-13: 978-1-57444-625-8 (hbk)
ISBN-13: 978-0-367-39302-1 (pbk)
Library of Congress Card Number 2005041703

Library of Congress Cataloging-in-Publication Data

Rekab, Kamel.
　　Statistical design of experiments with engineering applications / Kamel Rekab, Muzaffar Shaikh.
　　　　p. cm. -- (Statistics, textbooks and monographs ; v. 182)
　　Includes bibliographical references and index.
　　ISBN 1-57444-625-8 (alk. paper)
　　1. Experimental design. 2. Engineering—Statistical methods. I. Shaikh, Muzaffar. II. Title. III. Series.

QA279.R45 2005
519.5'7--dc22   2005041703

Visit the Taylor & Francis Web site at
http://www.taylorandfrancis.com

and the CRC Press Web site at
http://www.crcpress.com

Dedication

Dr. Rekab:

Dedicated to the memory of my father Lahcene and my youngest brother Toufik, my wife Josie, my daughters Sarah and Sonia, my mother Fettoum, and my brothers and sisters in Algeria.

Dr. Shaikh:

Dedicated in memory of my parents — A. Razzaque and Khadijah.

Acknowledgments

We would like to acknowledge the following persons whose help made this project come together: Khidir Abdelbasit, Obaid Al-Saidy, Ali Benmerzouga, Sergio Da Silva, Mark Fauls, Glenn Hess, Young-chan Lee, Eric Li, Dale Means, Mark Phelps, Thomas Sanders, Shoaib Shaikh, and Abootaleb Vafaie.

Preface

In today's high-technology world with flourishing e-Business and intense competition at a global level, the search for competitive advantage has become a crucial task of corporate executives. Corporations in the United States, Europe, and other parts of the world are incorporating scientific methods to accurately measure their performances to maintain or improve competitiveness. A key area of focus today is how to embed quality in products, services, processes, and systems. Quality, formerly thought secondary to expense, is now universally recognized as a necessary tool for success. Establishment, measurement, and maintenance of critical measures of performances (MOPs) are at the heart of sound quality engineering practices. And a key aspect of MOPs computation has to do with how they are measured. For example, if bias gets introduced, then the measured performance does not truly reflect the company's progress or lack of it. One important step that minimizes bias is the use of statistical methods in setting up designs of experiments that measure performances. Also, experiments need to be set up in a such a way that, on the one hand, they are cost-effective, and on the other, they do not lose key information. These experiments need to be efficient.

The authors have learned from years of experience with setting up experiments that a wide variety of experimental investigations needs to be tackled efficiently. Although a very large number of statistics books are available, experimenters are eager to have access to statistical methodologies that are easy to learn and to implement.

This book is intended for engineers, scientists, technicians, practitioners of scientific methods, and managers from a broad range of industries and disciplines. *Statistical Design of Experiments with Engineering Applications* is written for all of those who believe quality to be an entity that should be embedded in a product or service. Quality integration into a product or service has to be across all its life cycle stages. There are different tools and techniques that apply in different stages of the life cycle. In particular, one aspect of quality applies vividly in the design phase. It is very important to determine which factors affect the performance of the product or service design and to determine the settings of these factors that optimize the desired response.

The goal of this book is to elegantly implement common statistical tools and the new ones to improve the quality of the process that is used to manufacture a product or render a service. Throughout this book, we have used quality as the main theme to explain various design of experiments concepts. We firmly believe that quality can improve only by integrating it with experiments and field tests. Successful field tests of products and services ultimately result in high quality products. This book provides an applied approach to the design and analysis of experiments for students with little background in statistics. We have used this book for junior and senior level undergraduate engineering students and other closely related fields. We have also used part of this book in short courses and seminars for individuals with a wide variety of technical backgrounds from various industries. The feedback from all areas has been very favorable.

This book provides a ready-made and fast-learning approach to students and practitioners in applying design of experiments techniques to their problems without bogging them down in theoretical underpinnings. For example, if a practitioner has the need to design and analyze experiments with a view to optimize design parameters, this book is ideal in fulfilling such needs and avoids frustrating and unnecessary time spent on theory. Similar books by other publishers do not possess a clear-cut style for fast learning.

In a number of chapters, we provide stepwise procedures that need to be applied to industry problems. These steps are to the point and viable and they provide an easy vehicle to grasp a concept quickly and apply it to problems. Similar books in this area have avoided frequent use of stepwise procedures.

Where applicable, we provide practical rules of thumb for the practitioner to overcome certain potential problems while applying techniques. These rules of thumb are based on the vast experience of both authors in applied statistics. Books by other publishers do not possess these practical rules.

For each concept covered in each chapter we provide an example to clearly explain rather complicated concepts. This is very helpful to students and practitioners who want to apply techniques rather than spend time on learning the theory. Other books do provide examples but not in abundance.

Considering the importance of quality of products and services, we have attempted to include concepts from the quality engineering field throughout the book. Similar books by other publishers do not bring out the needed quality engineering angle in their texts.

The presentation of material in this book is arranged to include an introduction to the design of experiments (Chapter 1), a discussion of two-level design types (Chapter 2), a nonstatistical approach for location optimization (Chapter 3), a nonstatistical approach for variability reduction (Chapter 4), Taguchi's approach to the design of experiments (Chapter 5), a statistical approach for location optimization (Chapter 6), a statistical approach to variability reduction (Chapter 7), simple graphic methods and formal statistical tests for studying the validity of the prediction equation (Chapter 8), a discussion of three-level design types (Chapter 9), and an introduction to quadratic optimization (Chapter 10).

Table of Contents

CHAPTER 1

Introduction

1.1 WHAT IS EXPERIMENTAL DESIGN?

Experimental design deals with establishing and actually conducting a test or a series of tests in which changes are systematically made to the input variables of a process in order to observe the corresponding changes in the output response. The process is defined as a combination of machines, methods, people, and other resources that transforms some input into an output. For example, a fast-food franchise may want to observe the impact of changes in the quality of meat, toppings, buns, and equipment to determine the impact on the cost as the response variable.

The objective of experimental design is to provide the researcher or a practitioner with a statistical method that determines which input variables are most influential on the output and where to set the influential input variables so that the output is either maximized, minimized, or nearest to a desired target value. The design of the experiment approach can be applied to objectives that are as follows: smallest-is-the-best (e.g., cost), largest-is-the-best (e.g., profit), or nominal-is-the-best (e.g., room temperature at, say, 72°F). As an example, the field of agronomy, where experimental design was introduced, different fertilizer blends (input variables) were applied to crops in an effort to maximize crop yield (response).

One of the essential ideas underlying a designed experiment is that some methods of collecting input and output data are more powerful than others. In particular, the method of analyzing one input variable at a time, while holding the others fixed, turns out to be the least effective design. In a statistically designed experiment, the practitioner is able to change the much-needed and often-desired multiple variables to determine the impact of the response.

Though the design of experiments concept has long been used in the sciences, industry has not caught on with these methods since their introduction in the 1940s. In recent years, however, design of experiments has gained great popularity, primarily because of its great success in Japan where it was first introduced by W.E. Deming [1].

While Sir Ronald A. Fisher [2-4] was clearly the pioneer in the use of statistical methods in experimental design, there have been many other significant contributors to the literature of experimental design, including F. Yates [5-7], R.C. Bose [8-9], O. Kempthorne [10], W.G. Cochran [11-13], and G.E.P. Box [14-29], to name a few.

We now present several examples that illustrate some of the applications of the design of experiments.

1.2 APPLICATIONS OF EXPERIMENTAL DESIGN

Because of its nature, experimental design can be applied to many disciplines. In fact, it has been applied in manufacturing, finance, pharmaceutical companies, social sciences, biology, chemistry, and a multitude of other areas.

Example 1.1: *Minimizing a Process Parameter (Minimum-Is-the-Best)*
In an effort to minimize conversion as a response variable, four input variables were studied by a chemical engineer:
1. Catalyst charge $= 10 - 15$ lb
2. Temperature $= 220° - 240°$C
3. Pressure $= 50 - 80$ psi
4. Concentration $= 10\% - 12\%$

Which of the four variables are most influential on the conversion?
What settings should the engineer use to minimize the conversion?

Example 1.2: *Maximizing a Process Parameter (Maximum-Is-the-Best)*
A chemical engineer is interested in determining the operating conditions that maximize the yield of a process. Two variables were considered:
1. Reaction time $= 30 - 40$ minutes
2. Temperature $= 150° - 160°$F

Does the reaction time affect the yield?
Does temperature affect the yield?
What settings should the engineer use to maximize the yield?

Example 1.3: *Hitting a Target Parameter (Nominal-Is-the-Best)*
An engineer needs to improve turbine blade quality by reducing thickness variability around a target of 3 mm. Four variables were considered:
1. Metal temperature $= 20° - 22°$C
2. Mold temperature $= 3° - 5°$C
3. Pour time $= 1 - 3$ seconds
4. Vendor $=$ A and B

Which of the four variables affect the thickness?
What settings should the engineer use to meet the goal of 3 mm thickness?

Answers to these questions and others are addressed throughout this book.

Each of these examples can be represented by the model shown in Figure 1.1. For example, in the case of Example 1.3, there exist four input variables:

x_1 = metal pressure

x_2 = mold temperature

x_3 = pour time

x_4 = vendor

Figure 1.1. Model of a Process

The process may be a pressing operation and the output variable is thickness.

The examples that we provided can be further elaborated according to "the old philosophy of quality."

1.3 OLD PHILOSOPHY OF QUALITY

Historically, experienced engineers and scientists as experts in their respective fields have been known to establish specification limits — a lower specification limit (LSL) and an upper specification limit (USL) — and then perform inspections of finished products to ensure a minimum number of defects in products they studied. Loss to the company was based on being out of specification limits as shown in Figure 1.2 [30].

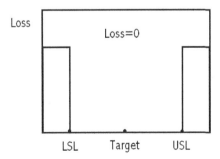

Figure 1.2. Philosophy of Quality

Under the old philosophy, corporations assume that as long as the product's response lies within the specification limits, no loss is incurred due to lack of quality. This traditional Goal-Post Loss function as shown in Figure 1.2 completely ignores losses even if a product or a process shifts from the target up to the specification limit on either side. Beyond the limits, losses jump up as a step function. Such a function representing losses is not realistic because losses start occurring as soon as the process shifts on either side of the target. It is also clear that there should be a great advantage for the product's response to be close to the target.

Professor Genichi Taguchi [30] introduced a different loss function that raises many important questions. The key component of Taguchi's philosophy is the reduction of variability.

In general, since the performance characteristic of each product or process has a target value, the objective is to reduce the variability around this target. For example, if a pen manufacturer sets a target of 150 mm for its length with specification limits, also termed as tolerance limits at ± 2 mm, then the losses begin even if the process variation is at, say, 151 mm or at 149 mm. Professor Taguchi imposes a loss function that will be described next. His loss function is the basis for the new philosophy of quality.

1.4 NEW PHILOSOPHY OF QUALITY

Taguchi advocates that for a product to be high in quality it must be on target and must have minimum deviation from the target. Under this new philosophy of quality, a loss is incurred as soon as a process shifts from the target. This is represented in Figure 1.3.

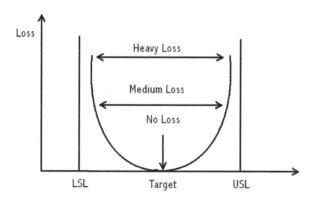

Figure 1.3. New Philosophy of Quality

As can be seen from Figure 1.3, if a process is right on target, there is no loss. As the process shifts from the target on either side, loss starts being incurred according to a parabolic function. The loss is at its maximum at the specification limits. It should also be noted that according to Taguchi, the loss function represents overall losses to the society. That is, the loss includes the actual loss to the firm as well as to the customer due to inconvenience or damages due to the lack of quality.

An interesting case of two products that average out to the same target value but have a significant difference in variability is illustrated in Figure 1.4.

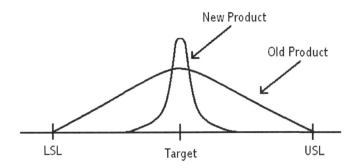

Figure 1.4. Minimum Variation around a Target Is Best

The old product averages to the target but has great variability as depicted by a large area under the curve. The new product, on the other hand, has small variability since the curve with smaller area hovers closely around the target.

We now present some additional examples that illustrate the new philosophy of quality.

Example 1.4: *Optimization of a Process with Minimum Variability*

To maximize the strength of a new material, an engineer considers two variables:

1. Temperature $= 200° - 240°C$
2. Pressure $= 50 - 80$ psi

Does the temperature affect the average strength?

Does the pressure affect the average strength?

Does the temperature affect the variability of the strength?

Does the pressure affect the variability of the strength?

What settings should the engineer use to maximize the strength with *minimum variability*?

Answers to these questions and others are addressed in the various chapters of this book.

1.5 ROBUST DESIGN

In all of the above examples, variables or factors, they are called in the design of experiments lexicon, reduce variability. As the name implies, values of controllable variables can be controlled by the experimenter. For example, in the experiment to maximize material strength (Example 1.4), the engineer has a thermostat to adjust the temperature to the desired value. In many experimental and field situations, some of the variables are controllable while others are uncontrollable. Uncontrollable factors are referred to as the noise. The effect of these factors cannot be controlled. Yet if the effect of the causes of variation can be minimized, then the resulting product or service will have minimum defects. Professor Taguchi introduced the concept of robust design. He provided a mechanism for finding the best settings that both reduce product variation and increase its robustness. A product is robust if it is resilient to variations in working environments. For example, shoes with special design features to waterproof them are robust in performance. The design in this case is robust.

The following example illustrates the need for robust designs.

Example 1.5: *Robust Design*

Three raw materials are used in varying amounts to form a certain plastic product. Depending on the quantities of each material used, the product will vary in strength, with the goal being to make the strongest product possible. It is also known that ambient temperature and humidity can affect the formation, and therefore the strength, of the resulting product.

What setting of the three raw materials should the engineer use so that the product's performance is robust against changes in temperature and humidity?

This example can be represented by a general model as shown in Figure 1.5.

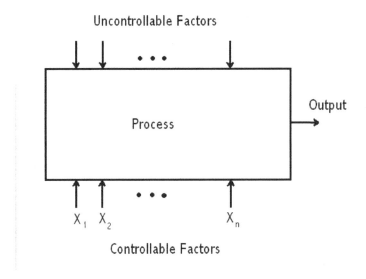

Figure 1.5. General Model of a Process or System

For Example 1.5, X_1 represents temperature, X_2 represents humidity, and output can be defined by, say, "brittleness" strength of the plastic. Uncontrollable factors would represent pure randomness or noises.

1.6 EXPERIMENTATION STEPS

Experimentation is integral to the design/redesign of a product, service, process, or system. All too often, ad hoc testing of a product or service makes it vulnerable to malfunctioning or a failure. For example, if a software module that involves tweaking of several variables to completely test all its functions is not tested for variations, all its variables may fail for certain combinations of their values. Therefore, it is imperative that experiments or tests are designed using a systematic approach. After presenting several examples and an introduction to the general design of experiments concept, we now present steps to sound experimentation. To achieve a successful experiment, four steps need to be followed: **planning** the experiment, **designing** the experiment, **conducting** the experiment, and **analyzing** the experiment.

Planning of an experiment consists of the following steps:

Step 1: Form the experimentation team. In forming the team, care must be taken to ensure that the team members are technically competent and are also easy to get along with. The team should be cohesive.

Step 2: Determine the objectives. An experiment without set objectives is doomed to failure. In the beginning, realistic objectives for the experiment must be established.

Step 3: Determine the response. This step deals with what output variable the experimenter wishes to study. This is a crucial step as it influences the next two steps.

Step 4: Select the input factors. This step deals with carefully determining the factors influencing the response. Omission of a factor or inclusion of a nonpertinent factor can impact the results in a negative way.

Step 5: Determine the settings of the factors. After establishing input variables, the experimenter needs to carefully establish the range of values of these variables for which tests will be conducted.

Designing the experiment consists of the following steps:

Step 1: Select the experimental design plan. In this crucial design step, the experimenter establishes the design plan that addresses such issues as how many observations or runs (i.e., the sample size), and affordability of the sample size.

Step 2: Assign factors to the experimental design plan. Once the design plan is established, the experimenter then can assign specific values of input variables under each observation.

Conducting the experiment consists of the following steps:

Step 1: Initial startup. This step in conducting tests includes a few trial runs and setting up of the experiment to remove any problems arising from initial runs. This permits smooth running of the actual experiment.

Step 2: Experimental runs. After the initial start up, runs are actually made in this step according to the design plan.

Step 3: Completion of experimental runs. This step ensures that all results are properly documented and the experimental setup and the equipment are brought back to their original states for future experimentation or use by others.

Analyzing the experiment consists of the following steps:

Step 1: Identify the important factors that affect the average response. In this step, input variables that have the largest influence on the outcome or response are identified.

Step 2: Find the optimal settings that minimize or maximize the average response. In this step, optimum values of input variables that minimize or maximize the average response are calculated.

Step 3: Identify the important factors that affect the variability. In this step, variables that have the greatest influence on the variability of the average response are identified.

Step 4: Find the best settings that minimize the variability. In this step, the best setting of variables that actually minimize the variability are established.

It should be noted that in all of the above steps, the engineer should employ suitable modeling and simulation tools depending upon the requirements of a particular problem. For example, in steps 1 and 4, the engineer may be able to formulate optimization models to determine optimum parameter values.

1.7 GOALS AND OUTLINE OF THE DESIGN OF EXPERIMENTS CONCEPTS

We conclude this section with a brief summary that describes the major goals of the Design of Experiments and present the necessary steps for performing a successful experiment.

A. Goals of the Design of Experiments

- Improve process yields (outcomes).
- Find factors that affect the average response (location).
- Find factors that affect the variability (dispersion).
- Find factor settings that optimize the average response (parameter optimization).
- Find factor settings that minimize variability.
- Reduce development time.
- Reduce overall costs.

B. Outline of the Design of Experiments

- Determine the objective of the experiment.
- Determine the response.
- Determine the factors and the region of interest.
- Determine the settings of the factors.
- Perform the experiment.
- Perform statistical analysis.
- Draw practical conclusions and give recommendations.

1.8 PROBLEMS

1. What is a designed experiment?
2. What is the difference between the old philosophy of quality and the new philosophy of quality?
3. What is the difference between noise factors and controllable factors?
4. What is a robust design?
5. Select a system or a process of your choice.
   a. Identify input variables, processing steps, and output variables for

this system or process.
b. Discuss how each input factor influences the output.

REFERENCES

[1] Deming, W.E. (1982), *Out of the Crisis*, Center for Advanced Engineering Study, MIT, Cambridge, MA.

[2] Fisher, R.A. and Yates, F. (1953), *Statistical Tables for Biological, Agricultural, and Medical Research*, 4th edition, Oliver and Boyd, Edinburgh.

[3] Fisher, R.A. (1958), *Statistical Methods for Research Workers*, 13th edition, Oliver and Boyd, Edinburgh.

[4] Fisher, R.A. (1966), *The Design of Experiments*, 8th edition, Hafner Publishing Co., New York.

[5] Yates, F. (1934), The analysis of multiple classifications with unequal numbers in the different classes, *J. of the Amer. Stat. Assoc.* **29**, 52-66.

[6] Yates, F. (1937), *Design and Analysis of Factorial Experiments*, Tech. Comm. No. **35**, Imperial Bureau of Soil Sciences, London.

[7] Yates, F. (1940), The recovery of interblock information in balanced incomplete block designs, *Annals of Eugenics* **10**, 317-325.

[8] Bose, R.C. and Shimamoto, T. (1952), Classification and analysis of partially balanced incomplete block designs with two associate classes, *J. of the Amer. Stat. Assoc.* **47**, 151-184.

[9] Bose, R.C., Clatworthy, W.H. and Shrikhande, S.S. (1954), Tables of partially balanced designs with two associate classes, *Tech. Bull. No.* **107**, North Carolina Agricultural Experimental Station.

[10] Kempthorne, O. (1952), *The Design and Analysis of Experiments*, Wiley, New York.

[11] Cochran, W.G. (1947), Some consequences when the assumptions for the analysis of variance are not satisfied, *Biometrics* **3**, 22-38.

[12] Cochran, W.G. (1957), Analysis of covariance: Its nature and uses, *Biometrics* **13**:3, 261-281.

[13] Cochran, W.G. and Cox, G.M. (1975), *Experimental Designs*, 2nd edition, Wiley, New York.

[14] Box, G.E.P. and Wilson, K.G. (1951), On the experimental attainment of optimum conditions, *J. of the Royal Statistical Soc.* **B**:13, 1-45.

[15] Box, G.E.P. (1954), Some theorems on quadratic forms applied to the study of analysis of variance problems I: Effect of inequality of variance in the one-way classification, *Annals of Math. Stats.* **25**, 290-302.

[16] Box, G.E.P. (1954), Some theorems on quadratic forms applied in the study of analysis of variance problems: II. Effect of inequality of variance and of correlation of errors in the two-way classification, *Annals of Math. Stats.* **25**, 484-498.

[17] Box, G.E.P. (1957), Evolutionary operation: A method for increasing industrial productivity, *Appl. Stats.* **6**, 81-101.

[18] Box, G.E.P. and Hunter, J.S. (1957), Multifactor experimental designs for exploring response surfaces, *Annals of Math. Stats.* **28**, 195-242.

[19] Box, G.E.P. and Behnken, D.W. (1960), Some new three level designs for the study of quantitative variables, *Technometrics* **2**, 455-476.

[20] Box, G.E.P. and Hunter, J.S. (1961), The 2^{k-p} fractional factorial designs, Part I, *Technometrics* **3**, 311-352.

[21] Box, G.E.P. and Hunter, J.S. (1961), The 2^{k-p} fractional factorial designs, Part II, *Technometrics* **3**, 449-458.

[22] Box, G.E.P. and Cox, D.R. (1964), An analysis of transformations, *J. of the Royal Stat. Soc.* **B**:26, 211-243.

[23] Box, G.E.P. and Draper, N.R. (1969), *Evolutionary Operation*, Wiley, New York.

[24] Box, G.E.P., Hunter, W.G. and Hunter, J.S. (1978), *Statistics for Experimenters*, Wiley, New York.

[25] Box, G.E.P. and Meyer, R.D. (1986), An analysis of unreplicated fractional factorials, *Technometrics* **28**, 11-18.

[26] Box, G.E.P. and Draper, N.R. (1987), *Empirical Model Building and Response Surfaces*, Wiley, New York.

[27] Box, G.E.P. (1988), Signal-to-noise ratios, performance criteria, and transformation, *Technometrics* **30**, 1-40.

[28] Box, G.E.P., Bisgaard, S. and Fung, C.A. (1988), An explanation and critque of Taguchi's contributions to quality engineering, *Quality and Reliability Eng. Intern.* **4**, 123-131.

[29] Box, J.F. (1978), *R.A. Fisher: The Life of a Scientist*, Wiley, New York.

[30] Taguchi, G. (1986), *Introduction to Quality Engineering*, Asian Productivity Organization, Hong Kong.

CHAPTER 2

Designing and Conducting the Experiment

2.1 INTRODUCTION

As mentioned in Chapter 1, it is extremely important that proper planning be done in first designing and then conducting experiments. The design engineer must adopt a proactive approach in setting up the design so that he/she is able to extract accurate information from the experiment. This ultimately leads to a product or service with minimal defects. On a new product, a well thought out design avoids rework during its development and implementation phase. It is with this motivation of objectivity in designs that we present some practical designs useful for the practicing engineer.

This chapter presents a wide variety of design types. These include one-factor-at-a-time design, two-level full factorial designs, fractional factorial designs of resolution III, Plackett-Burman designs, fractional factorial designs of resolution IV, and fractional factorial designs of resolution V. Advantages and disadvantages of each design type are also discussed to enable the engineer in deciphering which design to use for a particular application.

We begin with an example that shows how the one-factor-at-a-time design is very not efficient. Consequently, experimenters must use other types of designs that are cost efficient and easy to implement.

Example 2.1: An engineer is designing a battery for use in a device that will be subjected to extreme variations in temperature. The objective of the experiment is to maximize the lifetime of the battery. The engineer determined two factors that affect the lifetime of the battery:

1. Material type   O   N
2. Temperature   15°F   125°F

Now that the engineer has determined the objective of the experiment, influencing factors, their region of interest, and the response, the next logical steps are to set up the design, determine the settings of the factors, and perform the experiment.

2.2 ONE-FACTOR-AT-A-TIME APPROACH

Under this approach, the engineer varies only one factor at a time while keeping all other factors constant during an experiment. From a historical perspective, this type of design has been quite popular with engineers mainly due to its simplicity of use. For example, fix the material type to some level and vary the temperature. Take response measurements (i.e., life span of battery). Next fix the temperature and vary the material type. Again, take response measurements. The data collection is presented in Table 2.1.

Table 2.1. One-factor-at-a-time approach

Run	Material Type	Temperature	Life of Battery
1	O	15°	134.75
2	O	125°	57.50
3	N	125°	85.50

In runs 1 and 2, the material type was fixed at "O" and in run 3, material type was varied while the temperature was kept at 125. As it can be observed, the one-factor-at-a-time approach failed to examine an experimental run:

Material Type	Temperature	Life of Battery
N	15°	144

To avoid this disadvantage and to ensure that all extremities of input factors are addressed, factorial designs were introduced.

A full factorial design is a design in which all possible combinations of the levels of input factors are investigated. A full factorial design for the above example can be represented as in Table 2.2.

Table 2.2. Factorial design with two factors

Run	A	B	Y
1	+	+	85.50
2	−	+	57.50
3	+	−	144.00
4	−	−	134.75

Factor A represents the material type, factor B represents the temperature and Y represents the lifetime of the battery. A plus sign represents the maximum setting (second level), and a minus sign represents the minimum setting (first level). This representation is very useful as it will be shown in future chapters.

Another disadvantage of the one-factor-at-a-time design is its failure to examine the interaction between factors. For example, consider the following experiment:

Table 2.3. Two-level factorial design with interaction

Run	A	B	Y
1	−	−	20
2	+	−	50
3	−	+	40
4	+	+	12

To understand the factor interaction, we first need to understand the concept of main effect. The main effect of factor A is defined as the difference between the average response at the second level of A and the average response at the first level of A. That is,

$$\text{Main Effect of A} = \frac{50+12}{2} - \frac{20+40}{2} = 1.$$

This means that increasing factor A from its minimum level to maximum causes an average response increase of 1 unit, which is very small. This may lead us to conclude that there is no effect due to A. However, when examining the effects of factor A at different levels of factor B, we see that factor A is important. Figure 2.1 plots the response data in Table 2.3.

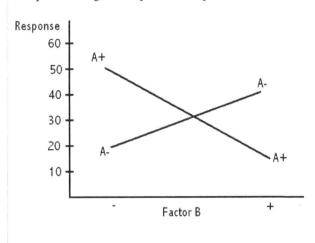

Figure 2.1. A Factorial Experiment with Interaction

At the lower level of B, the A effect is

$$A = 50 - 20 = 30$$

and at the higher level of B, the A effect is

$$A = 12 - 40 = -28.$$

Since the effect of A depends on the level chosen for factor B, we see that there is interaction between the factors. It is clear that A has an effect, but it depends on the level of factor B.

It is very important to determine the impact of interaction among factors. For example, in the pharmaceutical world, blood pressure medicine alone may not give a patient any headache, but if he/she starts taking simultaneously another medicine for some other sickness, he/she may start getting headaches.

Factorial designs can detect interaction between factors, and if the interaction is important, then the experimenter should not examine the factors separately.

On the other hand, suppose that the experimenter decided to perform a one-factor-at-a-time experiment as presented in Table 2.4.

Table 2.4. One-factor-at-a-time experiment

Run	A	B	Y
1	−	−	20
2	−	+	40
3	+	−	50

Table 2.4 indicates that if one factor is set at its high level and the other at its low, then the response is higher than the case where both factors are set at their low level. Values 40 and 50 are much higher than 20. It is very tempting to state that if both factors are set at their high level (i.e., at + and +) they will produce a greater response than $Y = 20$. However, if interaction is present, this conclusion may be wrong. As seen in Table 2.3, this value in fact is much lower at 12.

This clearly shows that factorial designs are more practical and informative than the one-factor-at-a-time experiment.

Next, we present some of the most useful factorial designs.

2.3 TWO-LEVEL FACTORIAL DESIGNS

In a two-level factorial design, each factor can only take on two values, for example, high and low. Henceforth, for the sake of convenience and without loss of generality, we set the low values at -1 and the high values at $+1$. Transformed variables -1 and $+1$ are called coded variables in the design of experiments context. Coded variables possess no units of measure. For example, consider factor A that takes values between 10 and 100. Figure 2.2 illustrates the relationship between coded variables and actual values.

Figure 2.2. Factor A Values and Their Transformations

We will map the low value to -1, and the high value to $+1$ using the following transformation:

$$z = \frac{x - [(\text{low} + \text{high})/2]}{[(\text{high} - \text{low})/2]},$$

where x denotes the actual value of factor A. For example, if $x = 10$ then $z = (10 - 55)/45 = -1$, if $x = 100$ then $z = (100 - 55)/45 = +1$ and if $x = 55$ then $z = (55 - 55)/45 = 0$.

Conversely, if a value of z is given, then the actual value of factor A can be calculated using:

$$x = z\frac{(\text{high} - \text{low})}{2} + \frac{(\text{high} + \text{low})}{2}.$$

For example, if $z = .25$ then $x = (.25)\left(\frac{100 - 10}{2}\right) + \frac{(100 + 10)}{2}$, which gives $x = 66.25$.

It should be noted that in an actual experiment, -1 and $+1$ values are established for other factors as well, similar to factor A.

A. Two-Level Full Factorial Designs

A two-level full factorial design with n factors requires 2^n experimental runs to cover all possible combinations of the input factors. This is illustrated in the factor columns in Table 2.5 for the case of two factors ($n = 2$) at two levels each. If there are five factors at $n = 5$, the total number of runs will be $2^5 = 32$.

Table 2.5. The 2^2 experimental design plan

Run	A	B	Response
1	+ 1	+ 1	y_1
2	− 1	+ 1	y_2
3	+ 1	− 1	y_3
4	− 1	− 1	y_4

We now present an example that illustrates an application of a two-level full factorial design.

Example 2.2:   Consider an investigation into the effect of the concentration of the reactant and the amount of the catalyst on the conversion (yield) in a chemical process. Let the reactant concentration be factor A and the two levels of interest be 15% and 25%. The catalyst is factor B, with the high level denoting the use of two bags of the catalyst and the low level denoting the use of only one bag.

The goal of this experiment is to answer the following questions.

1. Does concentration affect the yield?
2. Does catalyst affect the yield?

A two-level full factorial design for this problem would require at least four runs.

Following the experimental design plan of Table 2.5, the experimenter performed the experiment as follows:

Run	Concentration (A)	Catalyst (B)	Yield
1	25	2	30
2	15	2	20
3	25	1	33
4	15	1	26

The main effect of factor concentration is

$$A = \frac{30+33}{2} - \frac{20+26}{2} = 8.5,$$

and the main effect of factor catalyst is

$$B = \frac{30+20}{2} - \frac{33+26}{2} = -4.5.$$

The effect of concentration is positive; this suggests that increasing concentration from the lower level (15%) to the higher level (25%) will increase the yield.

The effect of the catalyst is negative; this suggests that increasing the amount of catalyst will decrease the yield.

A careful experimenter should investigate the interaction effect. Figure 2.3 plots the response data.

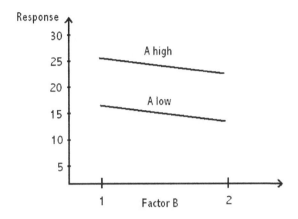

Figure 2.3. Very Low Interaction

Since the interaction is very small, the experimenter's decision about separating the effect of factor A and factor B is a good decision.

Many experimental design problems require more than two factors. It becomes very unclear how to analyze the interaction effect of factors if we want to rely on the response data.

Next, we will present a quantitative method to calculate the interaction between factors. To do so, the representation of Table 2.6 is necessary.

Table 2.6. Main effects and interaction effects

Run	A	B	AB	Y
1	+ 1	+ 1	+ 1	30
2	− 1	+ 1	− 1	20
3	+ 1	− 1	− 1	33
4	− 1	− 1	+ 1	26

The interaction effect is

$$\frac{30+26}{2} - \frac{20+33}{2} = 1.5,$$

which is smaller than the main effect of both factors A and B.

Determination of the Coded Value of an Interacting Variable: Referring to Table 2.6, the coded value of the interacting variable is calculated as follows:

Coded Value of AB = Coded Value of A × Coded Value of B.

For run 1, coded values of A and B are both $+1$. Therefore, the coded value of AB $= +1 \times +1 = +1$. We simply multiply algebraically the individual coded values of the two variables interacting together. Table 2.7 shows these calculations.

Table 2.7. Coded value of interaction variable

Run	A	B	AB
1	$+1$	$+1$	$(+1)(+1) = +1$
2	-1	$+1$	$(-1)(+1) = -1$
3	$+1$	-1	$(+1)(-1) = -1$
4	-1	-1	$(-1)(-1) = +1$

It should be noted that this multiplication scheme applies only to the coded values and not to the actual values of variables. For example, if A at $+1$ has a value of 100 and B at $+1$ has a value of 400, the AB value is not $100 \times 400 = 40000$. There are three reasons for why this multiplication is infeasible. First, units of measure of the two variables cannot be multiplied. Second, multiplied values are meaningless. Third, sometimes one of the variables A or B may be categorical and may not actually possess a value.

Next, an example with three factors is presented.

Example 2.3: An engineer is interested in the effects of cutting speed (A), tool geometry (B), and cutting angle (C) on the life (in hours) of a machine tool. Two levels of each factor are chosen, and a two-level full factorial design has been chosen.

A two-level full factorial design requires at least $2^3 = 8$ experimental runs.

Run	A	B	C	Y
1	−	−	−	26
2	+	−	−	35
3	−	+	−	40
4	+	+	−	50
5	−	−	+	42
6	+	−	+	38
7	−	+	+	57
8	+	+	+	42

The goal of this experiment is to determine which factors affect the life of the machine tool and to recommend the levels of the important factors that maximize the life of the machine tool.

The main effect of cutting speed is

$$A = \frac{35+50+38+42}{4} - \frac{26+40+42+57}{4} = 0,$$

the main effect of tool geometry is

$$B = \frac{40+50+57+42}{4} - \frac{26+35+42+38}{4} = 12,$$

and the main effect of cutting angle is

$$C = \frac{42+38+57+42}{4} - \frac{26+35+40+50}{4} = 7.$$

The effect of cutting speed is zero; this suggests that the cutting speed does not affect the life of the machine tool.

The effect of tool geometry is positive, thus suggesting that increasing the tool geometry from the low level to the high level will increase the life of the machine tool.

The effect of cutting angle is also positive, thus suggesting that increasing the cutting angle from the low level to the high level will increase the life of the machine tool.

It should also be noted that tool geometry has a greater influence at 12 on the tool life than the cutting angle at 7.

Next, a recommendation is to be made so that the life of the machine tool is maximized.

It is clear that both factors, tool geometry and cutting angle, should be set at their high levels. However, since the effect of the cutting speed is zero, we are tempted to conclude that it does not matter where to set the cutting speed.

A careful experimenter cannot stop here. He has to pay attention to the interaction between factors before giving a final recommendation.

Table 2.8 will be used to determine the two-factor interaction effects.

Table 2.8. Main effects and interaction effects

Run	A	B	C	AB	AC	BC	Y
1	−	−	−	+	+	+	26
2	+	−	−	−	−	+	35
3	−	+	−	−	+	−	40
4	+	+	−	+	−	−	50
5	−	−	+	+	−	−	42
6	+	−	+	−	+	−	38
7	−	+	+	−	−	+	57
8	+	+	+	+	+	+	42

The interaction effect between A and B is

$$AB = \frac{(26+50+42+42)}{4} - \frac{(35+40+38+57)}{4} = -2.5,$$

the interaction effect between A and C is

$$AC = \frac{(26+40+38+42)}{4} - \frac{(35+50+42+57)}{4} = -9.5,$$

and the interaction effect between B and C is

$$BC = \frac{(26+35+57+42)}{4} - \frac{(40+50+42+38)}{4} = -2.5.$$

Note that all two-factor interaction effects are negative, suggesting that setting AB, AC, and BC at -1 would increase the tool life.

Since the interaction effect between A and C is the largest, we need to analyze it more carefully. The effect of the AC interaction is negative; this suggests that a setting of AC at -1 would increase the life of the machine tool. Hence, the final recommendation to increase the life of the machine tool is to set the cutting speed at its low value, the tool geometry at its high value, and the cutting angle at its high value.

We are achieving a near-maximum tool life, since tool geometry and cutting angle are being set at high values providing positive impact on tool life (main effects of tool geometry $= 12$ and cutting edge $= 7$). Cutting speed at low combines with tool geometry and cutting angle, both at high to give a positive impact on tool life. Note that cutting speed and tool geometry interaction $= -2.5$ and that new cutting speed and cutting angle interaction $= -9.5$.

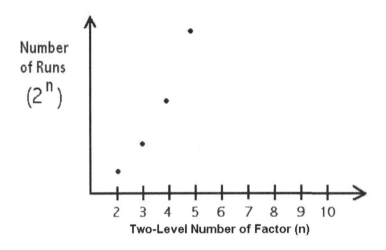

Figure 2.4. Number of Runs versus Number of Two-Level Functions

The objective of a two-level full factorial design is to test all possible combinations. Its advantage is evaluating all main and interaction effects, and its disadvantage is the large number of experimental runs. For example, if an experiment consists of studying 10 factors, a two-level full factorial design would require $2^{10} = 1024$ experimental runs.

As shown in Figure 2.4, number of runs increases dramatically when more and more factors are added. For most experiments it is practical to use a two-level full factorial design only when there are at most five factors. Beyond this, the experiment and analysis may become unmanageable.

Although the effects can always be calculated by using a table of plus and minus signs such as Table 2.7 or Table 2.8, it is difficult to carry on the calculations where n is large. An alternative approach is the tabular algorithm devised by Yates [1]. For a good description of this algorithm and an example, see Montgomery [2].

The biggest disadvantage of a two-level full factorial design is that the number of experimental runs increases exponentially with the number of factors as shown in Figure 2.4. From a cost standpoint, this can be very expensive. In some cases where it does not cost much to make extra runs, full factorial design may be acceptable. But when a single run costs a great deal (e.g., collecting water samples from a lake), then full factorial design becomes expensive. In such a case, an alternative needs to be devised. To partially alleviate this undesirable property, one may construct "fractional" factorial design plans by systematically eliminating some of the runs in a full factorial design.   For example, we wish to study the effect of three factors on a response. A two-level full factorial design requires $2^3 = 8$ experimental runs. A fractional factorial design requires only four experimental runs.

Table 2.9. Fractional factorial design with three factors

Run	A	B	C	Y
1	+	+	+	9
2	−	+	−	6
3	+	−	−	6
4	−	−	+	9

The main effect of factor A is

$$A=\frac{(9+6)}{2} - \frac{(9+6)}{2}=0,$$

the main effect of factor B is

$$B=\frac{(9+6)}{2} - \frac{(6+9)}{2}=0,$$

and the main effect of factor C is

$$C=\frac{(9+9)}{2} - \frac{(6+6)}{2}=3.$$

The effect of factor A is zero, suggesting that A does not affect the response; the effect of factor B is zero, suggesting that B does not affect the response; and the effect of C is positive, suggesting that increasing C from its low value to its high value will increase the response.

To determine the interaction effects between factors, we will use a representation similar to Table 2.7.

Run	A	B	AB	C	AC	BC	Y
1	+	+	+	+	+	+	9
2	−	+	−	−	+	−	6
3	+	−	−	−	−	+	6
4	−	−	+	+	−	−	9

The interaction effect between A and B is

$$\frac{(9+9)}{2} - \frac{(6+6)}{2}=3,$$

which is exactly the main effect of factor C; the interaction effect between A and C is

$$\frac{(9+6)}{2} - \frac{(6+9)}{2} = 0,$$

which is the same as the main effect of factor B; and the interaction effect between B and C is

$$\frac{(9+6)}{2} - \frac{(6+9)}{2} = 0,$$

which is the same as the main effect of factor A.

When this situation occurs, we say that A is confounded with BC, B is confounded with AC, and C is confounded with AB.

This phenomenon of, for example, "A is confounded with BC" etc. becomes very clear by simply observing the sequence of signs under A and under BC.

A	BC	Response
+	+	9
−	−	6
+	+	6
−	−	9

Obviously, both columns will result in exactly the same values.

Next we present the most useful two-level fractional factorial designs.

B. Fractional Factorial Designs of Resolution III

Fractional factorial designs of resolution III are designs in which no main effects are confounded with any other main effect. However, main effects are confounded with two-factor interactions and two-factor interactions may be confounded with each other. Fractional factorial designs were first introduced by Finney [3].

The construction of fractional factorial designs of resolution III is remarkably easy. We will proceed with a practical example.

Example 2.4: A statistical process control analyst is trying to determine which factors have a major effect in a circuit board etching process. The seven factors are as follows:

		Low	High
A:	Resist thickness	.1 mm	.5 mm
B:	Develop time	80 sec	90 sec
C:	Exposure	200	240
D:	Develop concentration	3.1:1	2.7:1

E:	Develop temperature	19°C	23°C
F:	Circuit line thickness	1 mm	3 mm
G:	Rinse time	5 sec	10 sec

A two-level full factorial design would require $2^7 = 128$ experiment runs. As discussed earlier, as more factors are added to an experiment, the number of runs increases exponentially. In this example, the analyst cannot afford that many runs. She, therefore, decides to investigate a fractional factorial design. A fractional factorial design of resolution III would require only 8 runs, thus saving 120 experimental runs. Now, we present the experimental design plan step by step.

Step 1: Determine the smallest integer that is a power of 2 and is larger than the number of factors. In Example 2.4, there are seven factors, the smallest integer which is a power of 2 and is larger than 7 is $8 = 2^3$. The experimental design plan requires $2^3 = 8$ runs.

Step 2: Construct a two-level full factorial design with three factors, where 3 is the power determined in step 1.

Runs	A	B	C
1	+	+	+
2	−	+	+
3	+	−	+
4	−	−	+
5	+	+	−
6	−	+	−
7	+	−	−
8	−	−	−

Step 3: Write out all of the two-factor interactions and the three-factor interactions. In Example 2.4, they are AB, AC, BC, and ABC.

Step 4: Confound any remaining main effect with a two-factor interaction or a three-factor interaction. In Example 2.4, D = AB, E = AC, F = BC, and G = ABC.

Step 5: Design the experiment with all factors. In Example 2.4, the analyst ran the following resolution III design.

Table 2.10. Fractional factorial design of resolution III with seven factors

Run	A	B	C	D(AB)	E(AC)	F(BC)	G(ABC)	Y
1	+	+	+	+	+	+	+	74
2	−	+	+	−	−	+	−	70
3	+	−	+	−	+	−	−	75
4	−	−	+	+	−	−	+	68
5	+	+	−	+	−	−	−	84
6	−	+	−	−	+	−	+	81
7	+	−	−	−	−	+	+	84
8	−	−	−	+	+	+	−	80

The main effect of resist thickness is

$$A = \frac{(74+75+84+84)}{4} - \frac{(70+68+81+80)}{4} = 4.5.$$

Similarly, main effects of all other variables are calculated and they are presented in Table 2.11 below.

Table 2.11. Main effects of input variables

Variable	Main Effect
A	4.5
B	0.5
C	− 10.5
D	− 1
E	1
F	0
G	− 0.5

These results clearly suggest that the resist thickness (A) and the exposure (C) have a significant effect on circuit board etching.

In Example 2.4 (Table 2.10), we have three *basic* factors A, B, and C. Relationships D = AB, etc. are called generating functions or generators. Note that original variables D, E, F, and G, which are not used in main effect calculations and remaining from the original list of seven variables (7 − 3 = 4), are called nonbasic factors. This means that there are four generators in this problem:

D = AB   E = AC
F = BC   G = ABC

All of the four generators were used in the experimental design plan proposed in Table 2.10. When this situation occurs, we say that the design is saturated.

We now present an artificial example that explains the construction of a fractional factorial design of resolution III.

Example 2.5: Suppose that 10 factors were chosen by an experimenter to determine which factors affect the response.

The construction of a fractional factorial design of resolution III uses the following stepwise procedure:

Step 1: The smallest integer that is a power of 2 and is larger than 10 is $16 = 2^4$. The experimental plan requires 16 runs.

Step 2: Construct a two-level full factorial design with four factors. There are four basic factors. We choose A, B, C, and D as our basic factors. The experimental plan with only A, B, C, and D is given in Table 2.12.

Table 2.12. Experimental plan with four variables

Run	A	B	C	D
1	+	+	+	+
2	−	+	+	+
3	+	−	+	+
4	−	−	+	+
5	+	+	−	+
6	−	+	−	+
7	+	−	−	+
8	−	−	−	+
9	+	+	+	−
10	−	+	+	−
11	+	−	+	−
12	−	−	+	−
13	+	+	−	−
14	−	+	−	−
15	+	−	−	−
16	−	−	−	−

Step 3: Write out all interaction terms of the four basic factors.

AB
AC
AD
BC
BD

CD
ABC
ABD
ACD
BCD
ABCD.

There are 11 generators and 6 nonbasic factors.

Step 4: We have 6 nonbasic factors E, F, G, H, J, and K that should be confounded with 6 interaction terms that are among the list of the 11 interaction terms. For example,

$E = ABC$
$F = ABD$
$G = ACD$
$H = BCD$
$J = ABCD$
$K = AB$

Usually, confounding is done with as many three-factor interactions as possible. This is because in most experiments, engineers are not interested in three-factor interactions. Instead, their interest centers around main effects or two-factor interactions. If confounding is done with two factors, then the engineer will not be able to differentiate whether the effect came from one factor or its corresponding two-factor interaction variables (e.g., $E = AB$). Therefore, in the above set of six confounding equations, five are with three-factor or higher interactions and one is two-factor. Since there are six nonbasic factors in this problem, we require at least six confounding equations.

Step 5: The fractional factorial design of resolution III with 10 factors is as follows:

Table 2.13. Resolution III with 10 factors

Run	A	B	C	D	E	F	G	H	J	K
1	+	+	+	+	+	+	+	+	+	+
2	−	+	+	+	−	−	−	+	−	−
3	+	−	+	+	−	−	+	−	−	−
4	−	−	+	+	+	+	−	−	+	+
5	+	+	−	+	−	+	−	−	−	+
6	−	+	−	+	+	−	+	−	+	−
7	+	−	−	+	+	−	−	+	+	−
8	−	−	−	+	−	+	+	+	−	+
9	+	+	+	−	+	−	−	−	−	+
10	−	+	+	−	−	+	+	−	+	−
11	+	−	+	−	−	+	−	+	+	−
12	−	−	+	−	+	−	+	+	−	+
13	+	+	−	−	−	−	+	+	+	+
14	−	+	−	−	+	+	−	+	−	−
15	+	−	−	−	+	+	+	−	−	−
16	−	−	−	−	−	−	−	−	+	+

The design above is usually denoted 2_{III}^{10-6}, where III is the type of resolution, 10 is the number of factors, and 6 is the number of generators that are used in the experimental design plan.

Resolution III Designs: Note that any column multiplied by itself results in a column of + values. This is because we are multiplying a + by + or a − by − , both resulting in + . A column that has only + values in each of its entries is denoted by I, the identity column.

In Example 2.4 and referring to Table 2.10, the following relations hold

    1.   ABD = I

    2.   ACE = I

    3.   BCF = I

    4.   ABCG = I

One obtains these identity relationships by simply starting with the first interacting variable (e.g., AB in Table 2.10) and multiplying it by its designated original variable (e.g., D) and proceeding this way until the end of the table is reached. In the literature, ABD, ACE, BCF, and ABCD are called words.

Note that the shortest word above has three letters. This is the reason why we refer to the design plan of Example 2.4 as a fractional factorial design of resolution III.

In Example 2.5, the following relations hold:

1.   ABCE = I
2.   ABDF = I
3.   ACDG = I
4.   BCDH = I
5.   ABCDJ = I
6.   ABK = I

Here again, the shortest word has three letters (i.e., ABK = I).

The objective of a fractional factorial design of resolution III is to test a fraction of a two-level full factorial design. Its advantage is that it saves many runs and its disadvantage is the loss of information about some two-factor interactions.

The following remarks are worth noting with respect to fractional factorial designs with resolution III.

Remark 1: Even though the construction of a fractional factorial design of resolution III is easy, most statisticians would rather refer to tables. See Appendix 1.

Remark 2: If the experimenter knows beforehand or assumes that the two-factor interactions are not important, then resolution III designs are appropriate to determine which factors are important.

Remark 3: Fractional factorial designs of resolution III require the number of experimental runs to be of a power of 2, for example 4, 8, 16, 32, 64, 128. Hence, there is no difference between the number of experimental runs for a design that has 32 factors or a design that has 63 factors; they both require the same number of experimental runs, that is, 64. This is because the smallest number larger than 32 that is also a power of 2 is 64 (2^8). For 63 factors, also, the largest number larger than 63 that is also a power of 2 is 64.

In view of Remark 3, Plackett and Burman [5] discovered a new type of design that requires in many cases less experimental runs than a fractional factorial design of resolution III.

Remark 4: Plackett-Burman designs do not contain information about the two-factor interactions.

Most engineers would rather use existing tables than construct them. In view of this, Taguchi [4] constructed sets of experimental design plans referred to as Taguchi designs (see Chapter 5). Next, we will present the two-level Plackett-Burman designs [5].

C.   Plackett-Burman (PB) Designs

These designs have the same objective as fractional factorial designs of resolution III. That is, they determine which factors are important. However, they have the advantage of requiring fewer experimental runs. Their disadvantage is their complexity. We have to rely on a single generating vector so that the experimental runs can be constructed. PB designs are explained as follows:

Let n denote the number of factors, and N_R denote the number of experimental runs. Plackett-Burman designs use a multiple of four as the number of experimental runs, that is, N_R = 8, 12, 16, 20, 24, 28, 32, 36, 40, etc. Also, the relationship between n and N_R is

$$N_R > n \text{ such that } N_R \text{ is a multiple of 4 starting with}$$

$$N_R = 8 \text{ as the smallest value.}$$

For example, suppose that the number of factors is 32; Plackett-Burman designs would use only 36 experimental runs, whereas fractional factorial designs of resolution III would use 64.

The following table compares the number of experimental runs of fractional factorial designs of resolution III with Plackett-Burman designs.

Table 2.14. Comparison between Plackett-Burman and 2_{III}^{n-p}

Number of factors	Plackett-Burman	Fractional factorial of resolution III
$4 \leq n \leq 7$	8	8
$8 \leq n \leq 11$	12	16
$12 \leq n \leq 15$	16	16
$16 \leq n \leq 19$	20	32
$20 \leq n \leq 23$	24	32
$24 \leq n \leq 27$	28	32
$28 \leq n \leq 31$	32	32
$32 \leq n \leq 35$	36	64
$\vdots$	$\vdots$	$\vdots$
$64 \leq n \leq 67$	68	128

Once the generating vector is determined, the construction of a Plackett-Burman design becomes relatively easy. We will proceed with an example.

Example 2.6: An experimenter needs to study the effect of 11 factors on the changes of an output response. He decides to perform experimental runs with a Plackett-Burman design. Only 12 experimental runs are needed. The construction of a Plackett-Burman design goes as follows:

Step 1: Get the single generating vector from Table 2.13 $n = 11$. The generating vector is:

$$+ \ + \ - \ + \ + \ + \ - \ - \ - \ + \ -$$

Step 2: Assign the generating vector to factor A and add a − for the last experimental run.

Run	Factor A
1	+
2	+
3	−
4	+
5	+
6	+
7	−
8	−
9	−
10	+
11	−
12	−

Step 3: Build the experimental runs for factor B by making the 11th value of factor A the first value of factor B, then slide the rest of factor A values below that value, and keep the last value of B at − .

Run	A	B
1	+	−
2	+	+
3	−	+
4	+	−
5	+	+
6	+	+
7	−	+
8	−	−
9	−	−
10	+	−
11	−	+
12	−	−

Step 4: Construct C by making the 11th value of factor B the first value of factor C, then slide the remaining values of factor B, and so on.

Table 2.15. Plackett-Burman 12 run design

Run	A	B	C	D	E	F	G	H	J	K	L
1	+	−	+	−	−	−	+	+	+	−	+
2	+	+	−	+	−	−	−	+	+	+	−
3	−	+	+	−	+	−	−	−	+	+	+
4	+	−	+	+	−	+	−	−	−	+	+
5	+	+	−	+	+	−	+	−	−	−	+
6	+	+	+	−	+	+	−	+	−	−	−
7	−	+	+	+	−	+	+	−	+	−	−
8	−	−	+	+	+	−	+	+	−	+	−
9	−	−	−	+	+	+	−	+	+	−	+
10	+	−	−	−	+	+	+	−	+	+	−
11	−	+	−	−	−	+	+	+	−	+	+
12	−	−	−	−	−	−	−	−	−	−	−

Next, the generating vectors for different numbers of factors are presented.

Table 2.16. Generators for Plackett-Burman designs

# Factors	# Runs	Generator
4-7	8	+ + + − + − −
8-11	12	+ + − + + + − − − + −
12-15	16	+ + + + − + − + + − − + − − −
16-19	20	+ + − − + + + + − + − + − − − − + + −
20-23	24	+ + + + + − + − + + − − + + − − + − + − − − −
32-35	36	− + − + + + − − − + + + + + − + + + − − + − − − − + − + − + + − − + −

We now present another example.

Example 2.7: An experimenter wishes to study the effect of seven factors on the changes of an output response. The experimenter decides to use a Plackett-Burman design. The PB construction will proceed as follows:

Step 1: Get the generating vector. Since $n = 7$, the Plackett-Burman design requires 8 runs, and the generator from Table 2.15 is

$$+ \ + \ + \ − \ + \ − \ −$$

Step 2: Assign the generating vector to factor A and add a − for the last entry.

Step 3: Build the experiment design plan by first making the 7th entry of A the first entry for B, then slide the remaining values of A, and keep the last value of B at − ; do the same for all factors.

Run	A	B	C	D	E	F	G
1	+	−	−	+	−	+	+
2	+	+	−	−	+	−	+
3	+	+	+	−	−	+	−
4	−	+	+	+	−	−	+
5	+	−	+	+	+	−	−
6	−	+	−	+	+	+	−
7	−	−	+	−	+	+	+
8	−	−	−	−	−	−	−

The following remarks are very important.

Remark 1: Even though the construction of Plackett-Burman designs is easy, most statisticians would rather refer to tables. See Appendix 2.

Remark 2: For many cases, Plackett-Burman designs and fractional factorial designs of resolution III are similar; in fact, they are equivalent when the number of experimental runs is expressed as a power of 2. For example, in Table 2.13, we have an equal number of runs when $12 \leq n \leq 15$ at 16.

Remark 3: Even though in many cases Plackett-Burman designs require fewer runs than fractional factorial designs of resolution III, they are more complicated to construct, and the confounding factors cannot be easily determined. Because of this complexity and the nonsystematic nature of the use of + and − levels, PB designs are good only for initial screening of factors in an experiment involving a large number of factors (e.g., 50 factors or more). Once an initial set of influencing factors is determined via PB designs, the experimenter can switch to, for example, resolution III fractional factorial or other systematic designs.

To alleviate the confounding between the main effects and the two-factor interactions, the engineer needs design types different from resolution III and PB. These designs are discussed in the next section.

D. Fractional Factorial Designs of Resolution IV

Fractional factorial designs of resolution IV are designs in which no main effect is confounded with any other main effect or two-factor interaction, but two-factor interactions are confounded with each other. The main advantage of these designs is that main effect measurements truly depict the impact of an input factor on the response, since it is not confounded with any other factor.

The construction of a fractional factorial design of resolution IV will be made through the following example.

Example 2.8: A human performance analyst is conducting an experiment to study the eye focus time and has built an apparatus in which several factors can be controlled during the test. The factors he initially regards as important are

A:	Sharpness of vision
B:	Distance from target to eye
C:	Target shape
D:	Illumination level
E:	Target size
F:	Target density
G:	Subject

He suspects that only a few of the seven factors are of major importance. On the basis of this assumption, the analyst decides to run a screening experiment to identify the most important factors and then to concentrate further study on those. He proceeds as follows:

Step 1: Construct a fractional factorial design of resolution III. The experimental design of resolution III was presented in Table 2.9. For completeness, we will present it again.

Run	A	B	C	D = AB	E = AC	F = BC	G = ABC	Time
1	+	+	+	+	+	+	+	141.8
2	−	+	+	−	−	+	−	95.0
3	+	−	+	−	+	−	−	77.6
4	−	−	+	+	−	−	+	83.7
5	+	+	−	+	−	−	−	145.4
6	−	+	−	−	+	−	+	93.2
7	+	−	−	−	−	+	+	75.1
8	−	−	−	+	+	+	−	85.5

The main effect of sharpness vision is

$$A = \frac{(141.8+77.6+145.4+75.1)}{4} - \frac{(95+83.7+93.2+85.5)}{4}$$
$$= 20.63,$$

and similarly,

B = 38.38
C = − 0.28
D = 28.88
E = − 0.28
F = − 0.63
G = − 2.43

As can be seen, the most important factors that affect the eye focus time are A, B, and D. We are tempted to conclude that sharpness of vision, distance from target to eye, and illumination level are the most important factors.

Since D = AB, then by multiplying this equality by B on both sides, BD = ABB = A. Note that BB is equal to I. Hence, A is confounded with BD. In fact, we can present all confounding patterns.

1.   A = BD = CE = FG
2.   B = AD = CD = EG
3.   C = AE = BF = DG
4.   D = AB = CG = EF
5.   E = AC = BG = DF
6.   F = BC = AG = DE
7.   G = CD = BE = AF.

Since C, E, F, and G are not important, then all interactions that include any of the four benign factors should also be considered not important. Hence, the only relations that we should investigate are

A = BD
B = AD
D = AB

Surprisingly, these three relations are all equivalent to ABD = I. Our conclusion that

(i)   A, B, and D are the most important could also mean
(ii)   A, B, and the AB interaction, or
(iii) B, D, and the BD interaction, or
(iv)   A, D, and the AD interaction are the three effects.

To determine with certainty which of the four statements is a correct one, the experimenter needs to add a second fraction with all the signs reversed. A design with signs reversed is called alternative half fraction or complementary half fraction.

Step 2: An additional fraction of eight experimental runs will be added. This additional fraction is similar to the fraction that was generated in step 1 except that all the signs will be reversed.

Run	A	B	C	D	E	F	G	Time
9	−	−	−	−	−	−	−	71.9
10	+	−	−	+	+	−	+	87.3
11	−	+	−	+	−	+	+	143.8
12	+	+	−	−	+	+	−	94.1
13	−	−	+	−	+	+	+	73.4
14	+	−	+	+	−	+	−	82.4
15	−	+	+	+	+	−	−	136.7
16	+	+	+	−	−	−	+	91.3

Step 3: Now, the analyst joins both fractions, the fraction that was determined in step 1 and the fraction that was determined in step 2. We obtain the following experimental design plan.

Run	A	B	C	D	E	F	G	Time
1	+	+	+	+	+	+	+	141.8
2	−	+	+	−	−	+	−	95.0
3	+	−	+	−	+	−	−	77.6
4	−	−	+	+	−	−	+	83.7
5	+	+	−	+	−	−	−	145.4
6	−	+	−	−	+	−	+	93.2
7	+	−	−	−	−	+	+	75.1
8	−	−	−	+	+	+	−	85.5
9	−	−	−	−	−	−	−	71.9
10	+	−	−	+	+	−	+	87.3
11	−	+	−	+	−	+	+	143.8
12	+	+	−	−	+	+	−	94.1
13	−	−	+	−	+	+	+	73.4
14	+	−	+	+	−	+	−	82.4
15	−	+	+	+	+	−	−	136.7
16	+	+	+	−	−	−	+	91.3

The generators for this design are
        E = BCD
        F = ACD
        G = ABC
An explanation of how these generators are obtained is as follows.

If we momentarily examine factors A, B, C, and D, they represent a full factorial design above with 16 runs. In step 1 above, we concluded that $A = BD$, $B = AD$, and $D = AB$ should be investigated. We can simply confound each two-factor interaction term with the remaining unused factor C of the full factorial portion of the overall design. This gives us (BD)(C), (AD)(C), and AB(C). Next, we can present the following relationships to complete the design:

$E = BCD$
$F = ACD$
$G = ABC$

The design above is usually denoted 2_{IV}^{7-3}, where IV is the resolution type, 7 is the number of factors, and 3 is the number of generators. The main effect of sharpness vision is

$$A = 1.48,$$

and similarly,

$B = 38.05$
$C = -1.80$
$D = 29.38$
$E = 0.13$
$F = 0.5$
$G = 0.13$

Now, it is clear to the analyst that distance from the target to eye (B) and illumination level (D) are the most important factors.

It should be noted that the main effects are confounded with the three level factor interactions. For example,

$A = CDF$
$B = ACG$
$C = ADF$
$D = BCE$
$E = BCD$
$F = ACD$
$G = ABC$

Most statisticians ignore the three-factor interactions because of their small impact. Hence, even though B is confounded with ACG, we are certain that the effect is due to B only.

We now present an artificial example that will reinforce our knowledge of the construction of a resolution IV design.

Example 2.9: Suppose that six factors were chosen by an experimenter to determine which factors affect the response. The construction of a fractional factorial design of resolution IV goes as follows:

Step 1: Construct a fractional factorial design of resolution III. We need eight experimental runs. We choose A, B, and C as our basic factors. The remaining three factors can be chosen from a set of the following generators:

AB
AC
BC
ABC

We choose

D = AB
E = AC
F = BC

The resulting resolution III design becomes

Run	A	B	C	D	E	F
1	+	+	+	+	+	+
2	−	+	+	−	−	+
3	+	−	+	−	+	−
4	−	−	+	+	−	−
5	+	+	−	+	−	−
6	−	+	−	−	+	−
7	+	−	−	−	−	+
8	−	−	−	+	+	+

Step 2: An additional fraction of eight experimental runs will be added where all of its signs are reversed.

Run	A	B	C	D	E	F
9	−	−	−	−	−	−
10	+	−	−	+	+	−
11	−	+	−	+	−	+
12	+	+	−	−	+	+
13	−	−	+	−	+	+
14	+	−	+	+	−	+
15	−	+	+	+	+	−
16	+	+	+	−	−	−

Step 3: Now, we join both fractions together to construct our design of resolution IV.

Run	A	B	C	D	E	F
1	+	+	+	+	+	+
2	−	+	+	−	−	+
3	+	−	+	−	+	−
4	−	−	+	+	−	−
5	+	+	−	+	−	−
6	−	+	−	−	+	−
7	+	−	−	−	−	+
8	−	−	−	+	+	+
9	−	−	−	−	−	−
10	+	−	−	+	+	−
11	−	+	−	+	−	+
12	+	+	−	−	+	+
13	−	−	+	−	+	+
14	+	−	+	+	−	+
15	−	+	+	+	+	−
16	+	+	+	−	−	−

The generators for this design are

$$E = BCD$$
$$F = ACD.$$

Note that we need only two generators here since we have a total of six variables and A,B,C and D result in full factorial measurements.

In some cases, it is more cost efficient to fold over a Plackett-Burman design. For example, suppose that in Example 2.6, the experimenter suspects that some of the two-factor interactions may be very important. The Plackett-Burman design with 12 experimental runs cannot be used because all Plackett-Burman designs are of resolution III. To alleviate this confounding problem, the experimenter can achieve a resolution IV design just by folding over the initial experimental plan of 12 runs.

This will result in the following experimental plan.

Table 2.17. Resolution IV by folding Plackett-Burman

Run	A	B	C	D	E	F	G	H	J	K	L
1	+	−	+	−	−	−	+	+	+	−	+
2	+	+	−	+	−	−	−	+	+	+	−
3	−	+	+	−	+	−	−	−	+	+	+
4	+	−	+	+	−	+	−	−	−	+	+
5	+	+	−	+	+	−	+	−	−	−	+
6	+	+	+	−	+	+	−	+	−	−	−
7	−	+	+	+	−	+	+	−	+	−	−
8	−	−	+	+	+	−	+	+	−	+	−
9	−	−	−	+	+	+	−	+	+	−	+
10	+	−	−	−	+	+	+	−	+	+	−
11	−	+	−	−	−	+	+	+	−	+	+
12	−	−	−	−	−	−	−	−	−	−	−
13	−	+	−	+	+	+	−	−	−	+	−
14	−	−	+	−	+	+	+	−	−	−	+
15	+	−	−	+	−	+	+	+	−	−	−
16	−	+	−	−	+	−	+	+	+	−	−
17	−	−	+	−	−	+	−	+	+	+	−
18	−	−	−	+	−	−	+	−	+	+	+
19	+	−	−	−	+	−	−	+	−	+	+
20	+	+	−	−	−	+	−	−	+	−	+
21	+	+	+	−	−	−	+	−	−	+	−
22	−	+	+	+	−	−	−	+	−	−	+
23	+	−	+	+	+	−	−	−	+	−	−
24	+	+	+	+	+	+	+	+	+	+	+

The experimenter should know in advance what is of interest to him/her. If the experimenter does not want to pay any attention to the two-factor interactions, probably because he knows that they are not important, then he will be well-advised to use any of the resolution III designs, either a 2_{III}^{n-p}, or a Plackett-Burman design, or even a Taguchi design. On the other hand, if the experimenter wants to pay attention to only a few two-factor interactions, then we would strongly advise a resolution IV design.

Next, we present an example that illustrates the above remark.

Example 2.10: To reduce the amount of shrinkage in a molded part, an experiment with four factors at two levels was performed. These are the factors:

A:	Set time	5 – 10 sec
B:	Zone 1 temperature	170° – 190° F
C:	Zone 2 temperature	170° – 190° F
D:	Preheat temperature	150° – 160° F

The experimenter does not anticipate any of the two-factor interactions, AD, BC, BD, and CD, to significantly impact the response.

The experimenter decides to construct a two-level fractional factorial design of resolution IV. This design requires eight runs, with three basic factors, A, B, and C, and one generator $D = ABC$.

The experimenter obtains the following response values:

Run	A	B	C	D	Y
1	+	+	+	+	5.3
2	−	+	+	−	6.2
3	+	−	+	−	4.2
4	−	−	+	+	6.9
5	+	+	−	−	3.8
6	−	+	−	+	8.8
7	+	−	−	+	7.3
8	−	−	−	−	6.6

The main effect of factor set time is $A = -1.975$, the main effect of factor zone 1 temperature is $B = -.225$, the main effect of factor zone 2 temperature is $C = -.975$, and the main effect of factor preheat temperature is $D = 1.875$.

From the main effects calculations, the experimenter has gained some very important information. That is, in order to reduce the amount of shrinkage, he should set all three factors — set time, zone 1 temperature, and zone 2 temperature — to their high levels, and he should set preheat temperature to its low level. Note that a negative main effect value that reduces the shrinkage implies increasing that factor will result in further shrinkage reduction.

Note also that this experimental run (+ + + −) has not been used in the experimental plan. And since A, B, and C are in significant shrinkage reduction, a careful experimenter should investigate the effect of the interactions AB and AC.

The generator $D = ABC$ enables the experimenter to determine the following confounding interaction terms:

1. $AB = CD$
2. $AC = BD$
3. $AD = BC$

The effect of the interaction AB or CD is $AB = -.975$. This is as significant as the main effect of factor C.

Hence, the experimenter should set AB at $+$ so that the amount of shrinkage decreases, and to do so, both factors A and B should be set at the same level (both high). Furthermore, even though AB and CD are confounded, the experimenter's assumption of the nonimportance of the interaction term CD enables him to be certain that the effect is due only to the interaction AB. The effect of the interaction AC or BD is AC $= .175$, which is smaller than any main effect.

In this example, as we just observed, the interaction between set time (A) and zone 1 temperature (B) was found to be important and thus should be considered as a potential factor in minimizing the amount of shrinkage. The experimenter's assumption about the nonimportance of the interaction between zone 2 temperature (C) and preheat temperature (D) enabled him with certainty to declare that the effect is due to AB only.

In many situations the experimenter does not know in advance anything about the two-factor interactions, especially when the design is new.

To alleviate the confounding between the two-factor interaction terms, other designs should be constructed.

The following remarks are very important.

Remark 1: Even though the construction of resolution IV designs is easy, most statisticians would rather refer to tables as a matter of convenience and to save time. Appendix 1 lists number of runs and generators for experiments involving number of factors and resolution.

Remark 2: Resolution IV designs are appropriate to determine with certainty which main effects are important, even in the presence of the two-factor interactions.

Remark 3: Even though in some cases it is cheaper to achieve a resolution IV design by folding over a Plackett-Burman design than to fold over a 2_{III}^{n-p} design, we prefer the fold over of a 2_{III}^{n-p} design. This is due to the fact that generators are very important to know. They determine the confounding elements.

Remark 4: In many situations, the experimenter does not know whether the two-factor interactions will be important. A design of resolution IV sometimes is not good enough because there is confounding between the two-factor interaction terms. In view of this, one should construct a design of resolution V.

E. Fractional Factorial Designs of Resolution V

Fractional factorial designs of resolution V are designs in which no main effect or two-factor interaction is confounded with any other main effect or two-factor interaction.

The construction of a resolution V design will be made through the following example.

Example 2.11: A researcher is investigating the effect of five factors on a chemical yield. The researcher concludes that the five factors are

1.	Feed rate	10 – 15 liters/min
2.	Catalyst	1% – 2%
3.	Agitation	100 – 120 rpm
4.	Temperature	140° – 180° C
5.	Concentration	3% – 6%

The investigator does not have any prior information about the two-factor interactions. He decides to construct an experimental plan of resolution V.

Step 1: Determine the number of experimental runs. In this case, a resolution III design requires eight experimental runs and a resolution V would require at least twice as many. Let us try to use 16 experimental runs.

Step 2: Construct a two-level full factorial design with only four factors.

Run	A	B	C	D
1	+	+	+	+
2	−	+	+	+
3	+	−	+	+
4	−	−	+	+
5	+	+	−	+
6	−	+	−	+
7	+	−	−	+
8	−	−	−	+
9	+	+	+	−
10	−	+	+	−
11	+	−	+	−
12	−	−	+	−
13	+	+	−	−
14	−	+	−	−
15	+	−	−	−
16	−	−	−	−

Step 3: Write all two-factor interactions, three-factor interactions, and four-factor interactions.

AB	BD	ABC	ABCD
AC	CD	ABD	
AD		ACD	
BC		BCD	

Step 4: Confound the remaining factor E with the highest interaction $E = ABCD$. With this chosen generator, no main effect or two-factor interaction is confounded with any other main effect or two-factor interaction. This can be seen throughout the following relations:

$A = BCDE$	$AB = CDE$
$B = ACDE$	$AB = BDE$
$C = ABDE$	$AD = BCE$
$D = ABCE$	$AE = BCD$
$E = ABCD$	$BC = ADE$
	$BD = ACE$
	$BE = ACD$
	$CD = ABE$
	$CE = ABD$
	$DE = ABC$

Note that

(i) $A = BCDE$ is obtained by multiplying both sides of the generator, $E = ABCD$, by BCD. That is, $(BCD)E = A(BCD(BCD) = A$.

(ii) The shortest word has five letters; that is, $ABCDE = I$.

Step 5: Now, the investigator decides to run the experiment.

Table 2.18. A 2_V^{5-1} design

Run	A	B	C	D	E	Yield (%)
1	+	+	+	+	+	82
2	−	+	+	+	−	95
3	+	−	+	+	−	60
4	−	−	+	+	+	49
5	+	+	−	+	−	93
6	−	+	−	+	+	78
7	+	−	−	+	+	45
8	−	−	−	+	−	69
9	+	+	+	−	−	61
10	−	+	+	−	+	67
11	+	−	+	−	+	55
12	−	−	+	−	−	53
13	+	+	−	−	+	65
14	−	+	−	−	−	63
15	+	−	−	−	−	53
16	−	−	−	−	+	56

The main effect of feed rate (A) is

$$A = -2.00,$$

the main effect of catalyst (B) is

$$B = 20.50,$$

the main effect of agitation rate (C) is

$$C = 0.00,$$

the main effect of temperature (D) is

$$D = 12.25,$$

and the main effect of concentration (E) is

$$E = -6.25.$$

It is clear that the catalyst (B), the temperature (D), and the concentration (E) are very important. It is also clear that agitation rate (C) has no effect on the yield, and the feed rate (A) is less important than the other three factors since it has a negative impact on the yield at $A = -2$.

Because of the signs of their main effects, by setting the catalyst at a high value, the temperature at a high value, and the concentration at a low value, the yield will certainly increase.

A further investigation needs to be made. Next, we will determine the effect of the two-factor interactions of the important factors. Simple calculations show that the only significant two-factor interactions are BD and DE.

The effect of the interaction BD is

$$BD = 10.75,$$

and the effect of the interaction DE is

$$DE = -9.50.$$

This justifies the fact that B and D should be set at the same level $(+)$, and the factors D and E should be set at the opposite levels.

Note that our conclusion about the settings of catalyst, temperature, and concentration is consistent with the experimental design plan.

Runs 2 and 5 are the only experimental runs that produce a percentage yield higher than 90%, and these two runs correspond to the following settings:

Catalyst (B) at $+$

Temperature (D) at $+$

Concentration (E) at $-$

Since the effect of agitation rate is null, it does not matter where to set it. If it is cheaper to set it at the low level, that would be fine. Even though its effect is not zero, we could attempt to set the feed rate at a low level.

Finally, we recommend

$$A\ B\ C\ D\ E$$
$$(-+-+-),$$

an experimental run that has not been used in the design plan.

We now present an additional example that should reinforce our knowledge of the construction of a resolution V design.

Example 2.12: A spin coater is used to apply photoresist to a bare silicon wafer. The coating thickness is of great importance. Six factors were identified as potential variables.

1. Final spin speed   A   6650 rpm 7300 rpm
2. Acceleration rate   B   5   20
3. Volume of resist applied C   3 cc   5cc
4. Time of spin   D   6 sec   14 sec
5. Resist batch variation   E   batch 1   batch 2
6. Exhaust pressure   F   cover off cover on

Construct a design of resolution V.

Step 1: Determine the number of experimental runs. In this case, if a resolution IV requires 16 runs, then a resolution V must require at least 32 runs.

Step 2: Construct a two-level full factorial design with five factors.

Run	A	B	C	D	E
1	+	+	+	+	+
2	−	+	+	+	+
3	+	−	+	+	+
4	−	−	+	+	+
5	+	+	−	+	+
6	−	+	−	+	+
7	+	−	−	+	+
8	−	−	−	+	+
9	+	+	+	−	+
10	−	+	+	−	+
11	+	−	+	−	+
12	−	−	+	−	+
13	+	+	−	−	+
14	−	+	−	−	+
15	+	−	−	−	+
16	−	−	−	−	+
17	+	+	+	+	−
18	−	+	+	+	−
19	+	−	+	+	−
20	−	−	+	+	−

21	+	+	−	+	−
22	−	+	−	+	−
23	+	−	−	+	−
24	−	−	−	+	−
25	+	+	+	−	−
26	−	+	+	−	−
27	+	−	+	−	−
28	−	−	+	−	−
29	+	+	−	−	−
30	−	+	−	−	−
31	+	−	−	−	−
32	−	−	−	−	−

Step 3: Write all two-factor interactions, three-factor interactions, four-factor interactions, and five-factor interactions:

AB	ABC	ABCD	ABCDE
AC	ABD	ABCE	
AD	ABE	ABDE	
AE	ACD	ACDE	
BC	ACE	BCDE	
BD	ADE		
BE	BCD		
CD	BCE		
CE	BDE		
DE	CDE		

Step 4: Confound the remaining factor F with the highest interaction,

$$F = ABCDE.$$

Step 5: Now, the experimenter can proceed by running the experiment.

Table 2.19. A 2_V^{6-1} design

Run	A	B	C	D	E	F
1	+	+	+	+	+	+
2	−	+	+	+	+	−
3	+	−	+	+	+	−
4	−	−	+	+	+	+
5	+	+	−	+	+	−

Table 2.19. (cont.)

Run	A	B	C	D	E	F
6	−	+	−	+	+	+
7	+	−	−	+	+	+
8	−	−	−	+	+	−
9	+	+	+	−	+	−
10	−	+	+	−	+	+
11	+	−	+	−	+	+
12	−	−	+	−	+	−
13	+	+	−	−	+	+
14	−	+	−	−	+	−
15	+	−	−	−	+	−
16	−	−	−	−	+	+
17	+	+	+	+	−	−
18	−	+	+	+	−	+
19	+	−	+	+	−	+
20	−	−	+	+	−	−
21	+	+	−	+	−	+
22	−	+	−	+	−	−
23	+	−	−	+	−	−
24	−	−	−	+	−	+
25	+	+	+	−	−	+
26	−	+	+	−	−	−
27	+	−	+	−	−	−
28	−	−	+	−	−	+
29	+	+	−	−	−	−
30	−	+	−	−	−	+
31	+	−	−	−	−	+
32	−	−	−	−	−	−

More details of fractional factorial designs are given in several experimental design books, such as those by Kempthorne [6], Davis [7], Hicks [8], Box, Hunter, and Hunter [9], Raktoe, Hedayat and Federer [10], Montgomery [2], and McLean and Anderson [11], as well as in the series of papers by Addelman [12-14], Margolin [15], and John [16-17].

The following remarks are very important.

Remark 1: Even though the construction of a resolution V design is easy, it is convenient to refer to the appendix for construction. See Appendix 1.

Remark 2: Resolution V is the most appropriate design to identify the importance of the main effects and the two-factor interactions.

2.4 PROBLEMS

1. List the disadvantages of one-factor-at-a-time experiments.
2. What is meant by a strong interaction between two factors?
3. Let A be a factor that takes its values between 100 and 1000.
   (a)   Compute the coded value (z) of x = 550.
   (b)   Compute the coded value (z) of x = 887.5.
   (c)   Compute the actual value (x) of z = 0.
   (d)   Compute the actual value (x) of z = .75.
4. Discuss the advantages/disadvantages of two-level full factorial designs.
5. What is meant by the following statement?
   A is confounded with B × C.
6. What is a design of resolution III?
7. When should you use a design of resolution III?
8. What is a design of resolution V?
9. When should you use a design of resolution V?
10. What is a saturated design?
11. What are the similarities/differences between fractional factorial designs and Plackett-Burman designs?
12. How do you obtain a resolution IV design from using a resolution III design?

REFERENCES

[1] Yates, F. (1937), *Design and Analysis of Factorial Experiments*, Tech. Comm. No. **35**, Imperial Bureau of Soil Sciences, London.
[2] Montgomery, D.C. (1991), Using fractional factorial designs for robust process development, *Quality Eng.* **3**, 193-205.
[3] Finney, D.J. (1945), The fractional replication of factorial arrangements, *Annals of Eugenics* **12**, 291-301.
[4] Taguchi, G. and Konishi, S. (1987), *Taguchi Methods: Orthogonal Arrays and Linear Graphs*, American Supplier Institute, Inc.
[5] Plackett, R.L. and Burman, J.P. (1946), The design of optimum multi-factorial experiments, *Biometrika* **33**, 305-325.
[6] Kempthorne, O. (1952), *The Design and Analysis of Experiments*, Wiley, New York.
[7] Davis, O.L. (1956), *The Design and Analysis of Industrial Experiments*, Oliver and Boyd, London.
[8] Hicks, C.R. (1982), *Fundamental Concepts of Design of Experiments*, 3rd edition, Holt, Rinehart, and Winston, New York.

[9] Box, G.E.P., Hunter, W.G. and Hunter, J.S. (1978), *Statistics for Experimenters: An Introduction to Design, Data Analysis and Model Building*, Wiley, New York.

[10] Ratkoe, B.L., Hedayat, A., and Federer, W.T. (1981), *Factorial Designs*, Wiley, New York.

[11] McLean, R.A. and Anderson, V.L. (1984), *Applied Factorial and Fractional Designs*, Marcel Dekker, New York.

[12] Addelman, S. (1961), Irregular fractions of the 2^n factorial experiments, *Technometrics* **3**, 479-496.

[13] Addelman, S. (1963), Techniques for constructing fractional replicate plans, *J. Amer. Statist. Assoc.* **58**, 45-71.

[14] Addelman, S. (1969), Sequences of two-level fractional factorial plans, *Technometrics* **11**, 477-509.

[15] Margolin, B.H. (1969), Results of factorial designs of resolution IV for the 2^n and $2^n 3^m$ series, *Technometrics* **11**, 431-444.

[16] John, P.W.M. (1961), The three-quarter replicates of 2^4 and 2^5 designs, *Biometrics* **17**, 319-321.

[17] John, P.W.M. (1962), Three-quarter replicates of 2^n designs, *Biometrics* **18**, 172-184.

CHAPTER 3

Optimization of the Location Parameter

3.1 INTRODUCTION

Every process contains at least one performance characteristic. We consider the case of a single performance characteristic, usually denoted by Y. For example, characteristics can be the yield of a process, the thickness of a turbine blade, the strength of a new material, the lifetime of a battery, the life of a machine tool, a circuit board etching, eye focus time, the amount of shrinkage in a molded part, a chemical yield, or the coating thickness of a silicon wafer.

The new philosophy of quality advocated by Taguchi states that a good product must be on target and must have minimum deviation from the target. Consequently, we should focus our interest on two parameters, the location parameter and the variation parameter. For instance, let us consider the following experiment.

Example 3.1: Consider an investigation into the effect of the concentration of a reactant and the amount of a catalyst on the conversion (yield) in a chemical process. Let the reactant concentration be factor A, and the two levels of interest be 15% and 25%. The catalyst is factor B, with the high level denoting the use of two bags and the low level denoting the use of only one bag. The experiment is replicated three times.

Table 3.1. The 2^2 experimental design plan

Run	A	B	Y_1	Y_2	Y_3
1	+	+	31	30	29
2	−	+	18	19	23
3	+	−	36	32	32
4	−	−	28	25	27

The location parameter is denoted by $\overline{Y}$; it is the average yield for a given combination of factors reactant concentration and the catalyst.

Thus, the location parameter in this experiment is represented by the following:

Table 3.2. Location parameter

Run	A	B	$\bar{Y}$
1	+	+	30.00
2	−	+	20.00
3	+	−	33.33
4	−	−	26.66

In this type of experimental scenario, the following questions become very important:

- Does concentration affect the location?
- Does catalyst affect the location?
- What settings of concentration and catalyst will result in maximizing the location?

In this chapter, we will focus on the location parameter, either maximizing it, minimizing it, or setting it to a target value. In the next chapter, we will discuss the variability of a process and ways of minimizing it. Next, we will present all the necessary steps that will result in optimizing the location parameter.

3.2 GUIDELINES FOR LOCATION OPTIMIZATION

1. Determine the characteristic of interest.
2. Determine the factors and the region of interest.
3. Determine the settings of the factors.
4. Construct a resolution V design.
5. Run the experiment.
6. Calculate the average response for every level.
7. Plot the averages.
8. Calculate the main effects and the two-factor interaction effects.
9. Generate a Pareto chart of the absolute value of the half effects.
10. Determine which factors and which two-factor interactions are important.
11. Generate a prediction equation using the important factors and the important two-factor interactions.
12. Based on the objective of the experiment, select the best settings of the factors to optimize the location parameter, and find the predicted optimum value of the location parameter.
13. Draw practical conclusions and give recommendations.

3.3 REPLICATED EXPERIMENTAL RUNS

To illustrate the 13 guidelines presented above, let us reconsider Example 3.1.

A. Maximizing the Location Parameter

1. The characteristic of interest is the conversion (yield).
2. The factors and the region of interest are
   - A: Concentration = 15% to 25%
   - B: Catalyst = 1 bag to 2 bags
3. The settings of the factors are:
   - A: 15% corresponds to − 1 (low)
   -          25% corresponds to + 1 (high)
   - B: 1 bag corresponds to − 1 (low)
   -          2 bags corresponds to + 1 (high)
4. Construct a resolution V design.

Run	A	B
1	+	+
2	−	+
3	+	−
4	−	−

5. Run the experiment.

Run	A	B	Y_1	Y_2	Y_3
1	+	+	31	30	29
2	−	+	18	19	23
3	+	−	36	32	32
4	−	−	28	25	27

6. Calculate the average response for every level.

Table 3.3. Complete experiment with average response

Run	A	B	AB	$\bar{Y}$
1	+	+	+	30.00
2	−	+	−	20.00
3	+	−	−	33.33
4	−	−	+	26.66

We first start by evaluating the average when A is set at its low level

$$\frac{20+26.66}{2} = 23.33,$$

and then, we calculate the average response when A is set at its high value

$$\frac{30+33.33}{2} = 31.67.$$

Similarly, for B and AB we obtain the following table:

Table 3.4. Average response per factor level

	A	B	AB
Low	23.33	30.00	26.67
High	31.67	25.00	28.33

7. Plot the averages.
   The plot of the averages is illustrated in Figure 3.1.

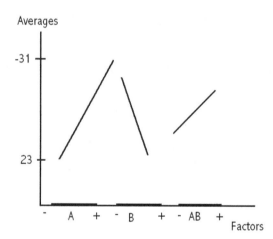

Figure 3.1. Plot of the Averages

This plot shows that concentration, catalyst, and the interaction contribute to the location parameter.
8. Calculate the main effects and the two-factor interaction effects.
   The main effect of concentration is

$$A = 31.67 - 23.33 = 8.34.$$

Similarly, the main effect of the catalyst and the two-factor interactions are presented in Table 3.5.

Table 3.5. Effects of A, B, and the AB interaction

	A	B	AB
effect=Δ	8.34	$-$ 5.00	1.66

9. Generate a Pareto chart of the absolute value of the half effects.
   The effect is denoted by Δ (delta), and the half effect is $\Delta/2$. Table 3.6 presents the absolute value of the half effects.

Table 3.6. Absolute value of the half effects

	A	B	AB
$|\Delta/2|$	4.17	2.50	.83

The Pareto chart is presented in Figure 3.2.

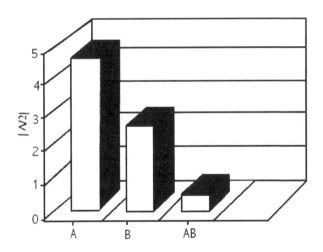

Figure 3.2. Pareto Chart of the Absolute Value of Half Effects

10. Determine which factors are important and which interactions are important.
    Figure 3.2 shows that both factors, concentration and catalyst, are important. It also shows that the two-factor interaction between concentration and catalyst is not important.

11. Generate a prediction equation using only the important factors and the important two-factor interactions.

The predicted location will be denoted $\hat{Y}$,

$$\hat{Y} = \overline{Y} + aA + bB,$$

where $\overline{Y}$ is the overall average; thus, $\overline{Y} = 27.50$, a is the half effect of factor concentration,

$$a = 4.17,$$

and b is the half effect of factor catalyst,

$$b = -2.50.$$

Most computer programs generate the regression coefficients using the method of least squares, which will be described in Chapter 10.

The prediction equation that relates the average yield to the important factors is

$$\hat{Y} = 27.50 + 4.17A - 2.50B.$$

12. Based on the objective of the experiment, select the best settings of the factors to optimize the location.

Here, the coefficient of concentration is positive and the coefficient of catalyst is negative.

If the objective is to maximize the yield, then factor A should be set at $+1$ and factor B should be set at -1.

The predicted maximum value of the conversion is

$$\hat{Y} = 27.50 + 4.17 \times (+1) - 2.5 \times (-1) = 34.17.$$

13. In order to maximize the yield, the concentration should be set at 25% with the use of only one bag of catalyst.

The analysis of this example agrees with the conclusion reached by the eyeball analysis approach, that is, by looking at the data and selecting the experimental run that maximizes the conversion. This is always the case with full factorial designs.

The reader may be wondering: Why do we need to go through the detailed work to create the prediction equation? There are three important reasons for this.

1. In many situations, full factorial designs are very expensive. In these situations, the prediction equation can be used to estimate the untested

combination on a fractional factorial design.
2. The equation is useful in predicting a target value.
3. The equation is used to perform sensitivity analysis.
Additionally, the prediction equation can also be used in the following two situations.

B. Prediction

The experimenter wishes to predict the yield of the process when the concentration is 22% and one and three quarter (1.75) bags of catalyst are used. The following approach is a general one and it can be used in all situations where prediction is of interest.
Recall the prediction equation

$$\hat{Y} = 27.5 + 4.17A - 2.50B.$$

First, we need to convert 22% into

$$\frac{22\% - (\text{low} + \text{high})/2}{(\text{high} - \text{low})/2},$$

which is equal to

$$\frac{22\% - (15\% + 25\%)/2}{(25\% - 15\%)/2} = .4.$$

Then, we need to convert 1.75 into

$$\frac{1.75 - (\text{low} + \text{high})/2}{(\text{high} - \text{low})/2},$$

which is equal to

$$\frac{1.75 - (1 + 2)/2}{(2 - 1)/2} = .5.$$

The predicted yield is obtained by substituting .4 into A and .5 into B. This yields

$$\hat{Y} = 27.50 + 4.17 \times .4 - 2.50 \times .5$$

$$= 27.50 + 1.67 - 1.25 = 27.92.$$

C. Hit a Target

The experimenter wishes to hit a target yield of 35 using only one bag of catalyst. What setting of concentration should he use to hit the target? The prediction equation is

$$\hat{Y} = 27.50 + 4.17A - 2.50B.$$

To obtain the catalyst, first, we convert 1 into

$$\frac{1-(1+2)/2}{(2-1)/2} = -1,$$

and then we substitute -1 into B and set $\hat{Y}$ to be 35. Thus,

$$35 = 27.50 + 4.17A + 2.50.$$

Now, we solve for A:

$$A = \frac{35-27.50-2.50}{4.17} = 1.2.$$

To determine the actual value of needed concentration, we use the following transformation:

$$(1.2) \times (\text{high} - \text{low})/2 + (\text{low} + \text{high})/2,$$

which is equal to
$$(1.2) \times (25\% - 15\%)/2 + (15\% + 25\%)/2$$

$$= (1.2) \times 5\% + 20\%$$

$$= 26\%.$$

Hence we need a concentration of 26% and one bag of catalyst to hit a target yield of 35.

We now present an additional example where the experimental runs are not replicated.

Example 3.2: Consider an investigation into the effect of five factors on a chemical yield. The factors are

1.	Feed rate	$10 - 15$ liters/min
2.	Catalyst	$1\% - 2\%$
3.	Agitation	$100 - 120$ rpm
4.	Temperature	$140° - 180°$ C
5.	Concentration	$3\% - 6\%$

In this type of experimental scenario, the following questions become important:

- Which factors are important?
- What settings of the factors will maximize the response?

We will follow the 13 guideline steps:

1. The characteristic of interest is the chemical yield.
2. The factors and the region of interest are

A:	Feed rate	10 – 15 liters/min
B:	Catalyst	1% – 2%
C:	Agitation	100 – 120 rpm
D:	Temperature	140° – 180° C
E:	Concentration	3% – 6%

3. The settings of the factors

A:	10 corresponds to – 1 (low)
	15 corresponds to + 1 (high)
B:	1 corresponds to – 1 (low)
	2 corresponds to + 1 (high)
C:	100 corresponds to – 1 (low)
	120 corresponds to + 1 (high)
D:	140 corresponds to – 1 (low)
	180 corresponds 59 + 1 (high)
E:	3 corresponds to – 1 (low)
	6 corresponds to + 1 (high)

4. Construct a resolution V design.

From Appendix 1, 16 runs are needed, with A, B, C, and D as the basic factors.

Table 3.7. Resolution V design

Run	A	B	C	D	E = ABCD
1	+	+	+	+	+
2	–	+	+	+	–
3	+	–	+	+	–
4	–	–	+	+	+
5	+	+	–	+	–

Table 3.7. (cont.)

Run	A	B	C	D	E = ABCD
6	−	+	−	+	+
7	+	−	−	+	+
8	−	−	−	+	−
9	+	+	+	−	−
10	−	+	+	−	+
11	+	−	+	−	+
12	−	−	+	−	−
13	+	+	−	−	+
14	−	+	−	−	−
15	+	−	−	−	−
16	−	−	−	−	+

5. Run the experiment.

Run	A	B	C	D	E	Y
1	+	+	+	+	+	82
2	−	+	+	+	−	95
3	+	−	+	+	−	60
4	−	−	+	+	+	49
5	+	+	−	+	−	93
6	−	+	−	+	+	78
7	+	−	−	+	+	45
8	−	−	−	+	−	69
9	+	+	+	−	−	61
10	−	+	+	−	+	67
11	+	−	+	−	+	55
12	−	−	+	−	−	53
13	+	+	−	−	+	65
14	−	+	−	−	−	63
15	+	−	−	−	−	53
16	−	−	−	−	+	56

6. Calculate the average response for every level.
    First, we calculate the average of main effects and interaction effects at the low level and then at the high level. We obtain the table below.

Table 3.8. Average responses per factor level

	A	B	C	D	E	AB	AC
Low	66.25	55.00	66.25	59.12	68.37	64.50	65.00
High	64.25	75.50	71.37	62.12	66.00	66.00	66.50

	AD	AE	BC	BD	BE	CD	CE	DE
Low	65.62	64.50	64.50	59.87	64.62	65.12	64.12	70.00
High	64.87	65.87	66.00	70.62	65.87	65.37	66.37	60.50

7. Plot the averages.
    The plot of the averages is illustrated below.

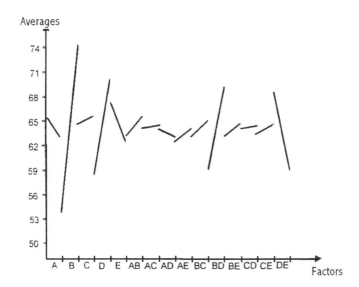

Figure 3.3. Plot of the Averages

This plot shows that the catalyst, the temperature, the concentration, the catalyst × temperature interaction, and the temperature × concentration interaction are important.
8. Calculate the main effects and the two-factor interaction effects.

Table 3.9. Main effects and two-factor interactions

	A	B	C	D	E
Δ	− 2.00	20.50	0.00	12.25	06.25

	AB	AC	AD	AE	BC	BD	BE	CD	CE	DE
Δ	1.50	0.50	− .75	1.25	1.50	10.75	1.25	0.25	2.25	− 9.50

9. Generate a Pareto chart of the absolute value of the half effects. Table 3.10 presents the absolute value of the half effects.

Table 3.10. Absolute value of half effects

	A	B	C	D	E	AB	AC
$|\Delta/2|$	1.00	10.25	0.00	6.125	3.125	0.75	0.25

	AD	AE	BC	BD	BE	CE	CE	DE
$|\Delta/2|$	0.375	0.625	0.75	5.375	0.625	0.125	1.125	4.75

The Pareto chart is presented in Figure 3.4.

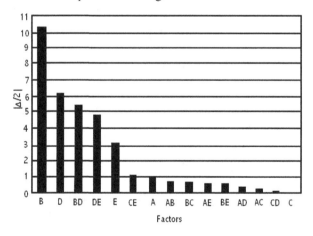

Figure 3.4. Pareto Chart of the Absolute Value of the Half Effects

10. Determine which factors and two-factor interactions are important.

   Both Figures 3.3 and 3.4 clearly indicate that the catalyst, the temperature, and the concentration are important factors; the other two-factor interactions that are important are catalyst × temperature and temperature × concentration.

11. Generate a prediction equation using only the important factors and the important two-factor interactions:

$$\widehat{Y} = \overline{Y} + aB + bD + cE + dBD + eDE,$$

$$\overline{Y} = 65.25.$$

a = 10.25 (half effect of catalyst)
b = 6.125 (half effect of temperature)
c = − 3.125 (half effect of concentration)
d = 5.375 (half effect of catalyst × temperature)
e = − 4.75 (half effect of temperature × concentration)
Y = 65.25 + 10.25B + 6.125D − 3.125E + 5.375BD − 4.75DE.

12. What settings of the factors should the experimenter use so that the yield is maximized? Factors B and D should be set at +1, and factor E should be set at − 1.

13. In order to maximize the yield, the experimenter should use 2% of catalyst, and he should set the temperature at 180°C, and the concentration at 3%. Since the factor agitation and feed rate are not important, their settings can be chosen for economical reasons.
The predicted maximum value of the yield is

$$\widehat{Y} = 65.25 + 10.25 \times (+ 1) + 6.125 \times (+ 1)$$

$$- 3.125 \times (- 1) + 5.375 \times (+ 1) \times (+ 1) - 4.75 \times (+ 1) \times (- 1)$$

$$= 95.$$

3.4 AN ALTERNATIVE APPROACH TO THE PARETO CHART

In determining which factors and two-factor interactions are important, we used the Pareto chart in previous exercises. We suggest a different approach that is somewhat similar to Sanders, Rekab et al. [1-2]. Further details can be found in Rekab and Sanders [3]. This approach will be explained through an example.

Suppose that A, B, and C are the factors used in an experimental study. Let

    a   denote the effect of A
    b   denote the effect of B
    c   denote the effect of C
    d   denote the effect of A × B interaction
    e   denote the effect of A × C interaction
    f   denote the effect of B × C interaction

Step 1: Evaluate $a^2 + b^2 + c^2 + d^2 + e^2 + f^2 = T$.

Step 2: Define the contribution of factor A as $\frac{a^2}{T} \times 100\%$ and calculate it. Repeat the same for B, C, AB, AC, and BC.

Step 3: Eliminate any factor or two-factor interaction whose contribution is less than a pre-established threshold value.

Step 4: Eliminate any two-factor interaction that has not been eliminated in step 3 if both factors have already been eliminated.

In Example 3.1, the contribution of concentration is 73%, the contribution of catalyst is 26%, and the contribution of the interaction is 1%.

Note that this approach shows us a much faster way to determine the important factors and the important two-factor interactions.

The following two examples will illustrate the new technique for determining, without any ambiguity, the important factors and the important two-factor interactions.

Example 3.3: An experimentation team wishes to maximize the strength of a woven textile. The team knows that the interaction between yarn type and any other factor is not important. It decided to follow our 13 guideline steps.

1. The characteristic of interest is the strength.
2. Factors and the region of interest are
   A: Side-to-side differences in strength (nozzle side of fabric versus opposite side)
   B: Yarn type (air spun versus ring spun)
   C: Pick density (the number of strands per unit inch; 35 strands versus 50 strands)
   D: Air pressure (30 versus 45 psi)
3. The settings of the factors are
   A: Nozzle corresponds to -1
      Opposite corresponds to $+1$
   B: Air spun corresponds to -1
      Ring spun corresponds to $+1$
   C: 35 strands corresponds to -1
      50 strands corresponds to $+1$
   D: 30 psi corresponds to -1
      45 psi corresponds to $+1$
4. Use a resolution IV design.

Table 3.11. Resolution IV design

Run	A	B	C	D
1	+	+	+	+
2	−	+	+	−
3	+	−	+	−
4	−	−	+	+
5	+	+	−	−
6	−	+	−	+
7	+	−	−	+
8	−	−	−	−

5. Conduct the experiment.

Table 3.12. Complete experiment

Run	A	B	C	D	Y_1	Y_2
1	+	+	+	+	24.00	24.46
2	−	+	+	−	25.36	24.00
3	+	−	+	−	24.02	25.00
4	−	−	+	+	31.36	20.00
5	+	+	−	−	24.00	26.00
6	−	+	−	+	25.04	24.00
7	+	−	−	+	24.10	20.00
8	−	−	−	−	24.00	25.00

6. Calculate the average response for every level.

Table 3.13. Complete experiment with average response

Run	A	B	C	D	$\overline{Y}$
1	+	+	+	+	24.23
2	−	+	+	−	24.68
3	+	−	+	−	24.51
4	−	−	+	+	25.68
5	+	+	−	−	25.00
6	−	+	−	+	24.52
7	+	−	−	+	22.05
8	−	−	−	−	24.50

Table 3.14. Average response per factor level

	A	B	C	D
Low	24.85	24.19	24.01	24.67
High	23.95	24.60	24.78	24.12

	AB	AC	AD	BC	BD	CD
Low	23.94	24.35	24.93	24.93	24.35	23.94
High	24.85	24.44	23.87	23.87	24.44	24.85

Note that
$$AB = CD$$
$$AC = BD$$
$$AD = BC$$
This is not surprising since the generator for our design is D=ABC.
7. Plot the averages.

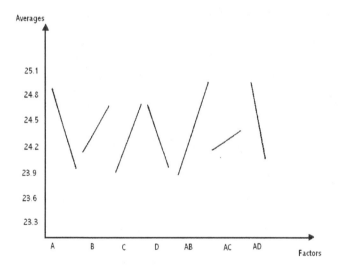

Figure 3.5. Plot of the Averages

8. Calculate the main effects and the two-factor interactions.

Table 3.15. Main effects and two-factor interactions

	A	B	C	D	AB	AC	AD
Δ	− .90	.41	.77	− .55	.91	.09	− 1.06

9. Generate a Pareto chart of the absolute value of the half effects.

Table 3.16. Absolute value of half effects

	A	B	C	D	AB	AC	AD
|Δ /2|	.450	.205	.385	.275	.455	.045	.530

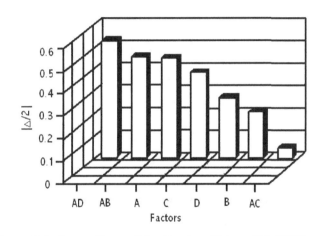

Figure 3.6. Pareto Chart of the Absolute Value of Half Effects

10. Determine which factors and two-factor interactions are important.

As can be seen from Figure 3.6, the Pareto chart failed to indicate which factors and two-factor interactions are important. In view of this, the alternative approach that was discussed at the beginning of this section will be used.

Let

$$a = \text{effect of A} = -.90$$
$$b = \text{effect of B} = .41$$
$$c = \text{effect of C} = .77$$
$$d = \text{effect of D} = -.55$$
$$e = \text{effect of A} \times B = .91$$
$$f = \text{effect of A} \times C = .09$$
$$g = \text{effect of A} \times D = -1.06$$

Now, the contribution of every factor and two-factor interaction may be evaluated.

Let $T = a^2 + b^2 + c^2 + d^2 + e^2 + f^2 + g^2 = 3.832$. The contribution of factor A is $\frac{(.90)^2}{3.832} = 21\%$, and contributions are calculated similarly for B, C, D, A × B, A × C, and A × D.

Table 3.17. Factor and two-factor interaction contribution

	A	B	C	D	AB	AC	AD
Contribution	21%	4%	15%	8%	22%	0%	30%

The team decided to use a threshold of 5%.

Since the contribution of factor B (yarn type) is less than 5%, B is not important. Since A × B = C × D, the contribution of 22% is due to the interaction C × D instead of A × B. The same argument can be used to show that the contribution of 30% is due to the interaction A × D instead of B × C. Furthermore, the contribution of A × C is negligible (almost 0%).

Therefore the remaining important factors and two-factor interactions are

$$A, C, D, A \times D, \text{ and } C \times D.$$

11. Generate a prediction equation using only the important factors and the important two-factor interactions.

The prediction average strength equation is

$$\hat{Y} = \bar{Y} + \tfrac{a}{2}A + \tfrac{c}{2}C + \tfrac{d}{2}D + \tfrac{e}{2}C \times D + \tfrac{g}{2}A \times D,$$

where $\bar{Y}$ is the overall average strength, so $\bar{Y} = 24.40$. Hence,

$$\hat{Y} = 24.40 - .45A + .385C - .275D - .53A \times D + .455C \times D.$$

12. Find the best settings to maximize the strength.

Since the order of the important factors and the two-factor interactions is

1st	AD
2nd	CD
3rd	A
4th	C
5th	D

AD should be set at − 1, which implies that A and D should be set at opposite levels. Since the coefficient of A is negative, A should be set at − 1; therefore, D will be set at + 1. Since the coefficient of CD is positive and D is set at + 1, C should be set at + 1.

Hence, the best setting to maximize the strength is the combination

$$(-?++).$$

Factor B could be set either at +1 or − 1. However, since it has a positive effect, B could be set at +1. Note that this combination (− +++) has not been used.

13. In conclusion, the most important factors that affect the strength are

Side-to-side, pick density, and air pressure.

Furthermore, to maximize the strength, the experimenter should set the side-to-side factor to nozzle side, the pick density should be set to 50 strands, the air pressure should be set at 45 psi, and yarn type could be set at ring spun.

The predicted maximum strength is

$$\hat{Y} = 24.40 - .45 \times (-1) + .385 \times (+1) - .275 \times (+1) - .53 \times (-1) \times (+1)$$

$$+ .455 \times (+1) \times (+1) = 25.945.$$

Another example of location optimization for unreplicated experimental runs and illustrating resolution IV design is given below.

Example 3.4: In an effort to minimize the amount of shrinkage in a molded part, an experimentation team decided to select four factors. The team decided to investigate all factors and only the interaction terms set time × zone 1 temperature and set time × zone 2 temperature. A resolution IV design is used instead of a more expensive design such as a resolution V design.

The team decided to follow our guidelines.

1. The characteristic of interest is amount of shrinkage.
2. The factors and their region of interest are

A:	Set time	5 − 10 sec
B:	Zone 1 temperature	170° − 190°F
C:	Zone 2 temperature	170° − 190°F
D:	Preheat temperature	150° − 160°F

3. Factor settings are

A:	5 corresponds to − 1
	10 corresponds to + 1
B:	170 corresponds to − 1
	190 corresponds to + 1
C:	170 corresponds to − 1
	190 corresponds to + 1

D: 150 corresponds to -1

160 corresponds to $+1$

4. A resolution IV design is constructed as follows:

Run	A	B	C	D
1	+	+	+	+
2	−	+	+	−
3	+	−	+	−
4	−	−	+	+
5	+	+	−	−
6	−	+	−	+
7	+	−	−	+
8	−	−	−	−

5. The experiment is conducted with the following result:

Table 3.18. Complete experiment

Run	A	B	C	D	Y
1	+	+	+	+	5.3
2	−	+	+	−	6.2
3	+	−	+	−	4.2
4	−	−	+	+	6.9
5	+	+	−	−	3.8
6	−	+	−	+	8.8
7	+	−	−	+	7.3
8	−	−	−	−	6.6

6. Next, average response for every level is calculated.

Table 3.19. Average response per factor level

	A	B	C	D	AB = CD	AC = BD	AD = BC
Low	7.125	6.250	6.625	5.200	6.625	6.050	5.925
High	5.150	6.025	5.650	7.075	5.650	6.225	6.350

7. Averages are plotted.

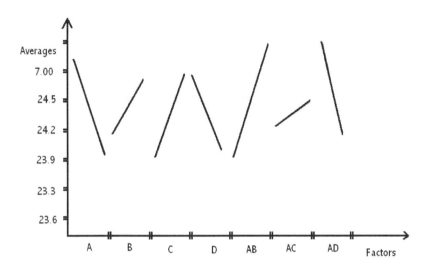

Figure 3.7. Plot of the Averages

This plot shows that factors A, C, and D are important factors, and also the interaction A × B is important.

8. Main effects and the two-factor interactions are calculated.

Table 3.20. Main effects and two-factor interactions

	A	B	C	D	AB(CD)	AC(BD)	AD(BC)
effect = Δ	− 1.975	− .225	− .975	1.875	− .975	.175	.425

9. A Pareto chart of the absolute value of the half effects is generated.

Table 3.21. Absolute value of half effects

	A	B	C	D	AB(CD)	AC(BD)	AD(BC)
|Δ/2|	.9875	.1125	.4875	.9375	.4875	.0875	.2125

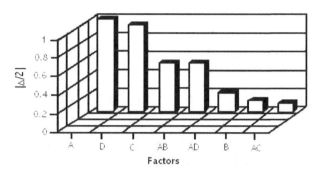

Figure 3.8. Pareto Chart of the Absolute Value of Half Effects

10.   The importance of single-factor and two-factor interactions is determined.

The plot of the averages and the Pareto chart of the absolute of the half effects clearly show that A, C, D, and AB are important.

We can also use the factor contribution technique to determine the important factors and the important two-factor interactions. This is accomplished as follows:

Let

$$a = \text{effect of A}$$
$$b = \text{effect of B}$$
$$c = \text{effect of C}$$
$$d = \text{effect of D}$$
$$e = \text{effect of AB(CD)}$$
$$f = \text{effect of AC(BD)}$$
$$g = \text{effect of AD(BC)}.$$

$T = a^2 + b^2 + c^2 + d^2 + e^2 + f^2 + g^2 = 9.5793$. The contribution of factor A is 41%, and, similarly, contributions of factors B, C, D, AB(CD), AC(BD), and AD(BC) can be calculated.

Table 3.22. Single-factor and two-factor interaction contribution

	A	B	C	D	AB(CD)	AC(BD)	AD(BC)
Contribution	41%	.5%	10%	37%	10%	.5%	1%

The team decided to use a threshold value of 5%.

Since the contribution of factor B (zone 1 temperature) is less than 5%, B is not important, and similarly, the interactions $A \times C$ and $A \times D$ are not important.

Therefore the remaining important factors and two-factor interactions are

$$A, C, D, \text{ and } A \times B.$$

11. A prediction equation using only the important factors and the important two-factor interactions is generated next.

$$\hat{Y} = \overline{Y} + \tfrac{a}{2}A + \tfrac{c}{2}C + \tfrac{d}{2}D + \tfrac{e}{2}AB$$

so

$$\hat{Y} = 6.1375 - .9875A - .4875C + .9375D - .4875A \times B.$$

12. Best settings to minimize the average amount of shrinkage are determined.

Since the coefficient of A is negative, A should be set at $+1$, C should be set at $+1$, and D should be set at -1. The coefficient of AB is negative, so $A \times B$ should be set at $+1$, and therefore B should be set at $+1$. Hence, the best settings to minimize the average amount of shrinkage are

$$(+ + + -).$$

13. In conclusion, the most important factors that affect the average amount of shrinkage are

Set time, preheat temperature, and zone 2 temperature.

Furthermore, in order to minimize the average amount of shrinkage, set time should be at 10 seconds, zone 1 and zone 2 temperatures should be both set at 190°F, and preheat temperature should be set at 150°F. Note that the experimental run $(+ + + -)$ has not been used.

The predicted minimum amount of shrinkage is

$$\hat{Y} = 6.1375 - .9875 \times (+1) - .4875 \times (-1)$$

$$+ .9375 \times (-1) - .4875 \times (+1) \times (+1) = 3.2375.$$

Before concluding this chapter, we give guidelines for choosing an appropriate design.

Guidelines for Design Choices

A. If all two-factor interactions are not important,
   - Use any design of resolution III, such as the two-level fractional factorial design 2_{III}^{n-p} or Plackett-Burman designs.
B. If only a few two-factor interactions are important,
   - Use any design of resolution IV, such as 2_{IV}^{n-p}, or folded Plackett-Burman design.
C. If all two-factor interactions are important,
   - Use a design of resolution V such as 2_{V}^{n-p} .

3.5 PROBLEMS

1. In chemical plants, the time to complete a particular filtration cycle is an important characteristic. Seven factors were selected for study:

A:	Water supply	town reservoir	well
B:	Raw material	on site	other
C:	Temperature	low	high
D:	Recycle	yes	no
E:	Caustic soda	fast	slow
F:	Filter cloth	new	old
G:	Hold up time	low	high

   The following experimental plan was chosen.

Filtration Time (min)

Run	A	B	C	D	E	F	G	Y
1	+	+	+	+	+	+	+	38.7
2	−	+	+	−	−	+	−	68.7
3	+	−	+	−	+	−	−	41.2
4	−	−	+	+	−	−	+	78.6
5	+	+	−	+	−	−	−	81.0
6	−	+	−	−	+	−	+	66.4
7	+	−	−	−	−	+	+	77.7
8	−	−	−	+	+	+	−	68.4
9	−	−	−	−	−	−	−	67.6
10	+	−	−	+	+	−	+	42.6
11	−	+	−	+	−	+	+	59.0
12	+	+	−	−	+	+	−	47.8
13	−	−	+	−	+	+	+	61.9
14	+	−	+	+	−	+	−	86.4
15	−	+	+	+	+	−	−	65.0
16	+	+	+	−	−	−	+	66.7

a. What type of design is this?
b. Present all confounding patterns.
c. What type of resolution is it?
d. Estimate the main effects and the two-factor interactions.
e. Which factors appear to be important in minimizing the amount of time to complete the filtration cycle?
f. Build a prediction model for the amount of time to complete the filtration cycle.
g. What are the optimal settings for minimizing the amount of time to complete the filtration cycle?

2. Redo questions a through e of problem 1 using only the first eight experimental runs.
3. Is it possible to answer questions f and g of problem 1 without any ambiguity if only the first eight experimental runs are used?

REFERENCES

[1] Sanders, T.J., Rekab, K., Rotella, F.M. and Means, D.P. (1992), Integrated circuit design for manufacturing through statistical simulation of process steps, *IEEE Trans. on Semiconductor Manufacturing* **4**, 368-372.

[2] Sanders, T.J., Rekab, K. and Chung, S.H. (1996), Statistical simulation of IC technology: A bipolar process example, *Microelectron. Reliab.* **36**, 1191-1205.

[3] Rekab, K. and Sanders, T.J. (1992), Variance estimation using full and fractional factorial designs, *Tech. Report No.* **1**, Department of Mathematics, Florida Institute of Technology, Melbourne, FL.

CHAPTER 4

Minimization of the Dispersion

4.1 INTRODUCTION

The Taguchi [1] loss function emphasizes the need to identify the factors that contribute to changes in the variability (dispersion). These factors are called dispersion factors.

In previous chapters, we focused our attention on the factors that contribute to changes in location. Before we study dispersion and its impact on experiments, consider the following graphs that illustrate four scenarios.

Scenario 1: Factors that affect the location but do not affect the dispersion can be represented as in Figure 4.1.

Distribution of Y when A is at —   Distribution of Y when A is at +

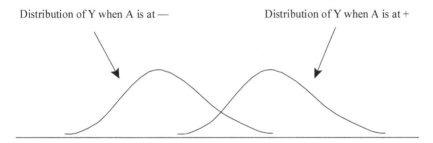

Figure 4.1. Factor A Affects the Location but Not the Dispersion

Scenario 2: Factors that affect the dispersion but do not affect the location can be represented as in Figure 4.2.

Distribution of Y when A is at —   Distribution of Y when A is at +

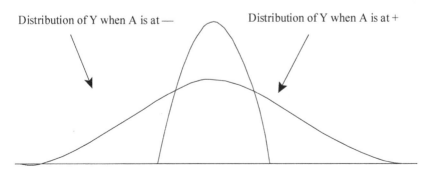

Figure 4.2. Factor A Affects the Dispersion but Not the Location

Scenario 3: Factors that affect both the location and the dispersion are represented in Figure 4.3.

Distribution of Y when A is at —   Distribution of Y when A is at +

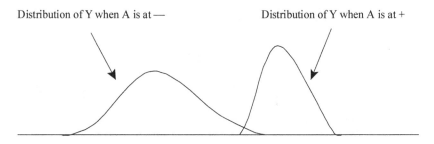

Figure 4.3. Factor A Affects Both Location and the Dispersion

Scenario 4: Factors that affect neither the location nor the dispersion are represented in Figure 4.4.

Both distributions are the same

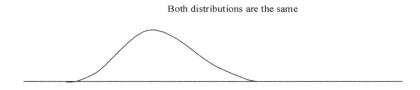

Figure 4.4. Factor A Affects Neither the Location nor the Dispersion

In this chapter, our attention will be focused on factors that affect the dispersion. Dispersion can be explained through a basic example. Suppose an experimenter wanted to compare two brands of batteries, A and B. He decided to use the following rule: A battery is good if it lasts for at least 625 hours. He took a sample of three A batteries and recorded their lifetimes: 500, 600, and 1000 hours. He then took a sample of three B batteries and recorded their lifetimes: 650, 700, 750 hours. It is clear that both brands have exactly the same average lifetime (700 hours). However, brand B is better because all three batteries lasted more than 625 hours. The main difference between the two brands is the dispersion. Brand A has a larger dispersion than brand B.

The goals of this chapter are to identify the important factors that affect the variability, to determine the settings of those factors that minimize the variability, and to predict the minimum variability.

Two sections are presented below to accomplish the above goal.

4.2 DISPERSION MINIMIZATION FOR REPLICATED STUDY

In order to explain the dispersion minimization for replicated studies, we start with an example.

Example 4.1: The object of the study is to minimize the yield variability. Factors are reactant concentration and catalyst, where concentration varies between 15% and 25%, and the catalyst levels are one bag for low and two bags for high.

Consider an investigation into the effect of the concentration of a reactant and the amount of a catalyst on the conversion (yield) in a chemical process. Let the reactant concentration be factor A, and the two levels of interest be 15% and 25%. The catalyst is factor B, with the high level denoting the use of two bags and the low level denoting the use of only one bag. The experiment is replicated three times.

Run	A	B	Y_1	Y_2	Y_3	$\bar{Y}$
1	+	+	34	30	29	31.00
2	−	+	18	19	23	20.00
3	+	−	36	32	32	33.33
4	−	−	28	25	27	26.66

The dispersion parameter is denoted by S. S is the sample standard deviation for a given combination of factors: reactant, concentration, and the catalyst.

The dispersion for the combination (+ +) is

$$S = \sqrt{\frac{(34-31)^2+(30-31)^2+(29-31)^2}{2}} = 2.65,$$

the dispersion for the combination (− +) is

$$S = \sqrt{\frac{(18-20)^2+(19-20)^2+(23-20)^2}{2}} = 2.65,$$

the dispersion for the combination (+ −) is

$$S = \sqrt{\frac{(36-33.33)^2+(32-33.33)^2+(32-33.33)^2}{2}} = 2.30,$$

and the dispersion for the last combination (− −) is

$$S = \sqrt{\frac{(28-26.66)^2 + (25-26.66)^2 + (27-26.66)^2}{2}} = 1.53.$$

Thus, the dispersion parameter in this experiment is represented by the following:

Table 4.1. Dispersion parameter

Run	A	B	S
1	+	+	2.65
2	−	+	2.65
3	+	−	2.30
4	−	−	1.53

In this study, the following questions are very important:
1. Does concentration affect the dispersion?
2. Does catalyst affect the dispersion?
3. What settings of concentration and catalyst will result in minimizing the dispersion?

Next, we present all the necessary guideline steps that result in minimizing the dispersion.

Guidelines for Dispersion Minimization Replicated Study

1. Determine the characteristic of interest.
2. Determine the factors and the region of interest.
3. Determine the settings of the factors.
4. Construct a resolution IV or V design.
5. Run the experiment.
6. Calculate the main effects and the two-factor interaction effects.
7. Determine which main factor effects and which two-factor interactions are important.
8. Generate a prediction equation that relates the dispersion with the important factors and the important two-factor interactions.
9. Select the best settings that minimize the dispersion.
10. Draw practical conclusions and give recommendations.

To illustrate the 10 guidelines, let us revisit Example 4.1.

Steps 1-5 are similar to steps 1-5 of the guidelines for the location parameter in Chapter 3.

Step 6: Calculate the main effects and the two-factor interaction effect.

Run	A	B	AB	S
1	+	+	+	2.65
2	−	+	−	2.65
3	+	−	−	2.30
4	−	−	+	1.53

The effect of concentration on the dispersion is

$$A = \frac{2.65+2.30}{2} - \frac{2.65+1.53}{2} = .385,$$

the effect of catalyst on the dispersion is

$$B = \frac{2.65+2.65}{2} - \frac{2.30+1.53}{2} = .735,$$

and the effect of concentration × catalyst is

$$AB = \frac{2.65+1.53}{2} - \frac{2.65+2.30}{2} = -.385.$$

Step 7: Determine which factors are important and which two-factor interactions are important.

We will proceed in a manner similar to the one we used in Section 3.3 of Chapter 3.

Let $T = (.385)^2 + (.735)^2 + (-.385)^2 = .837$. Then, the contribution of concentration to the dispersion is:

$$\frac{.148}{.837} \times 100\% = 18\%,$$

the contribution of catalyst to the dispersion is

$$\frac{.540}{.837} \times 100\% = 64\%,$$

and the contribution of catalyst × concentration is

$$\frac{.148}{.837} \times 100\% = 18\%.$$

Hence the catalyst, the concentration, and the interaction catalyst × concentration are all important.

Step 8: Determine the prediction equation

$$\hat{S} = \bar{S} + aA + bB + cAB,$$

where $\bar{S}$ is the average of the dispersions,

$$\overline{S} = \frac{(2.65+2.65+2.30+1.53)}{4} = 2.283,$$

a is the half effect of factor A, b is the half effect of factor B, and c is the half effect of the interaction AB. Then,

$$a = \frac{.385}{2} = .193,$$

$$b = \frac{.735}{2} = .368,$$

and

$$c = \frac{-.385}{2} = -.193.$$

Hence

$$\widehat{S} = 2.28 + .19A + .37B - .19AB.$$

Step 9: Select the best settings that minimize the dispersion. To minimize S, A, and B should be set at − 1 (low level).

Step 10: In order to minimize the dispersion of the yield, the concentration should be set at 15% with one bag of catalyst. The predicted minimum dispersion (variability) is

$$\widehat{S} = 2.28 + .19 \times (-1) + .37 \times (-1) - .19 \times (-1) \times (-1) = 1.53.$$

Next, we present an additional example that uses a fractional factorial design.

Example 4.2: A chemist wishes to minimize the variability of the yield response. Three factors are considered.

1. Temperature
2. Catalyst
3. pH

He decides to follow our guidelines for dispersion minimization for replicated study.

Step 1: The characteristic of interest is the yield variability (dispersion).

Step 2: Factors and the region of interest are

A:	Temperature	=	130° − 150°C
B:	Catalyst	=	1% − 2%
C:	pH	=	6.8 − 6.9

Step 3: Settings of the factors are

A: 130 corresponds to − 1 (low)
         150 corresponds to + 1 (high)

B: 1 corresponds to -1 (low)
            2 corresponds to $+1$ (high)
C: 6.8 corresponds to -1 (low)
            6.9 corresponds to $+1$ (high)

Step 4: The chemist does not anticipate any interaction effect. He then decides to proceed with a design of resolution III.

A	B	C
+	+	+
−	+	−
+	−	−
−	−	+

Step 5: Run the experiment.

Run	A	B	C	Y_1(week 1)	Y_2 (week 2)
1	+	+	+	75.3	77.1
2	−	+	−	61.2	59.6
3	+	−	−	75.4	73.1
4	−	−	+	66.0	63.3

Step 6: Calculate the main effects.
First, we shall determine the dispersion for each run.

Table 4.2. Dispersion parameter

Run	A	B	C	S
1	+	+	+	1.272
2	−	+	−	1.131
3	+	−	−	1.626
4	−	−	+	1.909

The effect of temperature on the dispersion is

$$A = \frac{1.272+1.626}{2} - \frac{1.131+1.909}{2} = -.071,$$

the effect of catalyst on the dispersion is

$$B = \frac{1.272+1.131}{2} - \frac{1.626+1.909}{2} = -.566,$$

and the effect of pH on the dispersion is

$$C = \frac{1.272+1.909}{2} - \frac{1.131+1.626}{2} = .212.$$

Step 7: Determine which factors are important (affect the dispersion the most).

Let $T = (.071)^2 + (.566)^2 + (.212)^2 = .370$. The contribution of temperature on the dispersion is

$$\frac{.005}{.370} \times 100\% = 1\%,$$

the contribution of catalyst on the dispersion is

$$\frac{.320}{.370} \times 100\% = 87\%,$$

and the contribution of pH on the dispersion is

$$\frac{.045}{.370} \times 100\% = 12\%.$$

This shows that catalyst and pH are the most important factors that affect the dispersion.

Step 8: The prediction equation is

$$\widehat{S} = \overline{S} + bB + cC,$$

where $\overline{S}$ is the average dispersion, b is the half effect of catalyst, and c is the half effect of pH.

Thus

$$\widehat{S} = 1.484 - .283B + .106C.$$

Step 9: Select the best settings to minimize the yield variability.

Since the coefficient of B is negative, B should be set at $+1$, and since the coefficient of C is positive, C should be set at -1.

Even though A is not important, but since its effect is negative, A will be set at $+1$. Thus the experimental run that minimizes the yield variability is

$$(+ + -).$$

Note that this run has not been used.

Step 10: In order to minimize the yield variability, the catalyst should be set at 2%, and pH should be set at 6.8. The predicted minimum dispersion (variability) is

$$\widehat{S} = 1.484 - .283 \times (+1) + .106 \times (-1) = 1.095.$$

Dispersion minimization for replicated study can also be achieved through other statistical methods. Bartlett and Kendall [2] proposed the use of log S^2 as the predictor instead of S. Montgomery [3] showed that the use of $\ln\frac{S_+^2}{S_-^2}$, where S_+^2 is the sample variance when the factor is set at its high level and S_-^2 is the sample variance when the factor is set at its low level, is also a sound approach for minimizing the dispersion. Further details for modeling variance are discussed by Carroll and Ruppert [4].

Now we explain the dispersion minimization when studies are unreplicated.

4.3   DISPERSION MINIMIZATION FOR UNREPLICATED STUDY

Let us consider the data in Example 3.2 where the object of the study is to minimize the dispersion of a chemical yield.

Example 4.3.   Consider the experimental design plan of Example 3.2, where

A:	Feed rate	10 – 15 liters/min
B:	Catalyst	1% – 2%
C:	Agitation	100 – 120 rpm
D:	Temperature	140° – 180°C
E:	Concentration	3% – 6%

Y is a chemical yield.

Run	A	B	C	D	E	Y
1	+	+	+	+	+	82
2	−	+	+	+	−	95
3	+	−	+	+	−	60
4	−	−	+	+	+	49
5	+	+	−	+	−	93
6	−	+	−	+	+	78
7	+	−	−	+	+	45
8	−	−	−	+	−	69
9	+	+	+	−	−	61
10	−	+	+	−	+	67
11	+	−	+	−	+	55
12	−	−	+	−	−	53
13	+	+	−	−	+	65
14	−	+	−	−	−	63
15	+	−	−	−	−	53
16	−	−	−	−	+	56

Questions we wish to address are
- Which factors affect the dispersion?
- What settings of the factors will result in minimizing the dispersion?
  We will present all necessary steps to minimize the dispersion.

Guidelines for Dispersion Minimization Unreplicated Study

1. Determine the characteristic of interest.
2. Determine the factors and their regions of interest.
3. Determine the settings of the factors.
4. Construct a resolution IV or V design.
5. Run the experiment.
6. Generate a prediction equation that relates the location and all important factors and two-factor interactions.
7. Calculate the residuals $e = Y - \hat{Y}$.
8. Determine which factors affect the dispersion.
9. Select the best settings that minimize the dispersion.
10. Draw practical conclusions and give recommendations.
    To illustrate the 10 guidelines, let us consider Example 4.3.
    Steps 1-5 are similar to steps 1-5 of the guidelines for the replicated study.
    Step 6: Generate a prediction equation that relates the location and all important factors and two-factor interactions.
    From Example 3.2, the prediction equation that relates the yield and the important factors is

$$\hat{Y} = 65.25 + 10.25B + 6.125D - 3.125E + 5.375BD - 4.75DE \,.$$

Step 7: Calculate the residuals.
    In run #1, the observed yield is $Y = 82$ and the predicted yield is calculated by substituting $+1$ for A, B, C, D, and E, so

$$\hat{Y} = 65.25 + 10.25 + 6.125 - 3.125 + 5.375 - 4.75 = 79.12.$$

The residual for the first run is

$$e = 82 - 79.12 = 2.88.$$

We will do the same calculations for all the remaining 15 runs.

Table 4.3. Residuals for unreplicated study

Run	A	B	C	D	E	Y	$\hat{Y}$	e
1	+	+	+	+	+	82	79.12	2.88
2	−	+	+	+	−	95	94.87	.13
3	+	−	+	+	−	60	63.62	− 3.62
4	−	−	+	+	+	49	47.87	1.13
5	+	+	−	+	−	93	94.87	− 1.87
6	−	+	−	+	+	78	79.12	− 1.12
7	+	−	−	+	+	45	47.87	− 2.87
8	−	−	−	+	−	69	63.62	5.38
9	+	+	+	−	−	61	62.37	− 1.37
10	−	+	+	−	+	67	65.62	1.38
11	+	−	+	−	+	55	55.87	− .87
12	−	−	+	−	−	53	52.62	.38
13	+	+	−	−	+	65	65.62	− .62
14	−	+	−	−	−	63	62.37	.63
15	+	−	−	−	−	53	52.62	.38
16	−	−	−	−	+	56	55.87	.13

Step 8: Determine which factors affect the dispersion. This will require four steps.

Step 8.1: For every factor and two-factor interaction, evaluate the average of the residuals at each level.

When factor A is set at its high level, the average of the residuals is

$$\frac{2.88 - 3.62 - 1.87 - 2.87 - 1.37 - .87 - .62 + .38}{8} = -.995.$$

Step 8.2: For every factor and two-factor interaction, evaluate the variance of the residuals at each level.

When factor A is set at its high level, the variance of the residuals is

$$[(2.88 - (-.995))^2 + (-3.62 - (-.995))^2 + (-1.87 - (-.995))^2$$

$$+ (-2.87 - (-.995))^2 + (-1.37 - (-.995))^2 + (-.87 - (-.995))^2$$

$$+ (-.62 - (-.995))^2 + (.38 - (-.995))^2]/7$$

$$= \frac{28.37}{7} = 4.05,$$

and similar calculations will be performed for the low level.

The table of variances for every factor and two-factor interaction is shown below.

Table 4.4. Variance for every factor and two-factor interaction level

variance	A	B	C	D	E	AB	AC
low	2.76	7.51	6.12	.803	6.879	3.194	2.160
high	4.05	2.50	3.91	9.23	3.160	5.696	7.732

variance	AD	AE	BC	BD	BE	CD	CE	DE
low	4.786	2.857	2.337	8.018	1.857	6.642	2.059	6.946
high	5.072	6.286	6.410	2.018	7.205	3.571	5.000	3.089

Step 8.3: For every factor and two-factor interaction, evaluate $\ln \frac{S_+^2}{S_-^2}$ where ln is the natural logarithm, S_+^2 is the sample variance of the residuals when the factor is set at its high level, and S_-^2 is the sample variance of the residuals when the factor is set at its low level.

For example, the dispersion effect of factor A is

$$\ln \frac{S_+^2}{S_-^2} = \ln \frac{4.05}{2.76} = .383,$$

and similar calculations will be performed for the remaining factors and the two-factor interactions. The table below presents the natural logarithm of the ratio of variances.

Table 4.5. Natural logarithm of ratio of variances

	A	B	C	D	E	AB	AC	AD	AE
$\ln \frac{S_+^2}{S_-^2}$	.383	-1.099	$-.448$	2.441	$-.778$	.578	1.274	.058	.788

	BC	BD	BE	CD	CE	DE
$\ln \frac{S_+^2}{S_-^2}$	1.009	-1.379	1.356	$-.620$	.887	$-.810$

Step 8.4: If the absolute value of dispersion effect exceeds 1.96, we conclude that the dispersion effect is real.

Here we see that factor D (the temperature) is the only factor that affects the dispersion.

Step 9: To minimize the dispersion, factor D should be set at its low level $(-)$.

Step 10: In order to minimize the dispersion (variability), the temperature should be set at 140°C. The other factors do not affect the dispersion and thus they can be set either at the low or at the high level.

The statistical approach presented in steps 6, 7, and 8 was developed by Box and Meyer [5].

Another example of dispersion minimization for unreplicated experimental runs is given below.

Example 4.4: Consider the experimental design plan of Example 3.4, where

A:	Set time	5 – 10 sec
B:	Zone 1 temperature	170° – 190°F
C:	Zone 2 temperature	170° – 190°F
D:	Preheat temperature	150° – 160°F

The response Y is the amount of shrinkage.

Run	A	B	C	D	Y
1	+	+	+	+	5.3
2	−	+	+	−	6.2
3	+	−	+	−	4.2
4	−	−	+	+	6.9
5	+	+	−	−	3.8
6	−	+	−	+	8.8
7	+	−	−	+	7.3
8	−	−	−	−	6.6

- Which factors affect the dispersion?
- What settings of the factors will result in minimizing the dispersion?

Steps 1-5 are similar to those of Example 3.4.

Step 6: Generate a prediction equation that relates the location and all important factors and two-factor interactions.

From Example 3.4, the prediction equation that relates the amount of shrinkage and the important factors is

$$\hat{Y} = 6.137 - .987A - .487C + .937D - .4875A \times B.$$

Step 7: Calculate the residuals $e = Y - \hat{Y}$.

In run #1, the observed amount of shrinkage is $Y = 5.3$ and the predicted amount of shrinkage is calculated by substituting $+1$ for A, B, C and D, so

$$\hat{Y} = 6.137 - .987 - .487 + .937 - .4875 = 5.11.$$

The residual for this first experimental run is

$$e = Y - \hat{Y} = 5.3 - 5.11 = .19.$$

We will do the same for all remaining seven runs.

Table 4.6. Residuals for unreplicated runs

Run	A	B	C	D	Y	$\hat{Y}$	e
1	+	+	+	+	5.3	5.11	.19
2	−	+	+	−	6.2	6.19	.01
3	+	−	+	−	4.2	4.21	− .01
4	−	−	+	+	6.9	7.08	− .18
5	+	+	−	−	3.8	4.21	− .41
6	−	+	−	+	8.8	9.03	− .23
7	+	−	−	+	7.3	7.06	.24
8	−	−	−	−	6.6	6.19	.41

Step 8: Determine which factors affect the dispersion. This will require four steps.

Step 8.1: For every factor and two-factor interaction, evaluate the average of the residuals at each level.

When A is set at its low, the average of the residuals at that level is

$$\frac{.01 - .18 - .23 + .41}{4} = .0025.$$

Step 8.2: For every factor and two-factor interaction, evaluate the variance of the residuals at each level.

When A is set at its low level, the variance of the residuals is

$$[(.01 - .0025)^2 + (-.18 - .0025)^2 + (-.23 - .0025)^2 + (.41 - .0025)^2]/3$$

$$= .25/3 = .08.$$

The table of variances of every factor and two-factor interaction level is shown below.

Table 4.7. Variances per factor and two-factor interaction level

Variance	A	B	C	D	AB	AC	AD
low	.08	.07	.15	.11	.04	.08	.03
high	.09	.07	.02	.06	.13	.07	.03

Step 8.3: For every factor and two-factor interaction, evaluate $\ln\frac{S_+^2}{S_-^2}$, where ln is the natural logarithm, S_+^2 is the variance when the factor is at its high level, and S_-^2 is the variance when the factor is at its low level.

For example, for factor A,

$$\ln\frac{S_+^2}{S_-^2} = .12,$$

and we will repeat the same calculations for all the factor and the two-factor interactions. The table below presents the natural logarithm of the ratio of the variances.

Table 4.8. Natural logarithm of the ratio of variances

	A	B	C	D	AB	AC	AD
$\ln\frac{S_+^2}{S_-^2}$	.12	.00	-2.01	$-.61$	1.30	$-.02$	.00

Step 8.4: If the absolute value of dispersion effect exceeds 1.96, we conclude that the dispersion effect is real. Here, only factor C (zone 2 temperature) affects the dispersion.

Step 9: To minimize the variability of the amount of shrinkage, factor C should be set at its high level (+).

Step 10: In order to minimize the dispersion (variability) of the amount of shrinkage, zone 2 temperature should be set at 190°F. The other factors do not affect the dispersion, so they can be set to any level.

Before concluding this chapter, guidelines for choosing an appropriate design are given. These guidelines are similar to those given in Chapter 3. They are included here simply for completeness.

Guidelines for Design Choices

A. If all two-factor interactions are not important,
   - Use any design of resolution III, such as two-level
     fractional 2_{III}^{n-p} factorial design or Plackett-Burman designs.
B. If only a few two-factor interactions are important,
   - Use any design of resolution IV, such as 2_{IV}^{n-p}, or folded
     Plackett-Burman design.
C. If all two-factor interactions may be important,
   - Use a design of resolution V such as 2_V^{n-p}.

4.4 PROBLEMS

1. Consider the experimental design plan of problem 1 from Chapter 3.
   a. Which factors affect the dispersion of the amount of time to complete the filtration cycle?
   b. What are the optimal settings for minimizing the dispersion of the amount of time to complete the filtration cycle?
   c. What are the optimal settings for minimizing the amount of time to complete the filtration cycle while maintaining the dispersion at its minimum?

REFERENCES

[1] Taguchi, G. (1986), *Introduction to Quality Engineering*, Asian Productivity Organization, Hong Kong.

[2] Barlett, M.S. and Kendall, D.G. (1946), The statistical analysis of variance heterogeneity and the logarithmic transformation, *J. of the Royal Statist. Soc.* Series **B**:8, 128-150.

[3] Montgomery, D.C. (1991), *Design and Analysis of Experiments*, 3rd edition, Wiley, New York.

[4] Caroll, R.J. and Ruppert, D. (1988), *Transformation and Weighting in Regression*, Chapman & Hall, New York.

[5] Box, G.E.P. and Meyer, R.D. (1986), Dispersion effects from fractional designs, *Technometrics* **28**, 19-27.

CHAPTER 5

Taguchi's Approach to the Design of Experiments

5.1 INTRODUCTION

Before the 1980s, American engineers had little knowledge on how to use statistical methods to improve processes. They were using the best guess and one-factor-at-a-time methods. Even though a very large number of statistical methods were available, these methods were complicated. Statisticians could not cross the bridge in finding applications of their sophisticated theoretical results to engineering. And many of the methods such as factorial designs and optimization, among others, were created in the West. Professor Taguchi was able to modify some of the existing statistical tools and made them easy enough to understand so that engineers could apply them. Even though there are a large number of statisticians who tend to disagree with most of Taguchi's contributions, there are many engineers who have reported great successes by using his modified statistical methods.

To name one of his success stories, Bell Telephone Laboratories in New Jersey, USA, and another electrical communications laboratory based in Japan, where Taguchi held a position, were developing similar cross-bar telephone exchange systems. In addition to facing Bell's greater resources, the Japanese were constrained by inferior materials. Despite all the adversity, the new cross-bar system from Japan was rated superior and cost much less to produce. The effect was so dramatic that Western Electric stopped production and began importing systems from Japan.

The emphasis of Dr. Taguchi's methodology is on functional variation, which can be measured in terms of product performance such as strength, pressure, shrinkage, response time, taste, and mean time to failure, among many others. Viewed as the enemy of the producer and its customer, functional variation can relate to the performance of the end product or to the process that manufactures the end result.

The purpose of experimentation using Taguchi methodology is to identify those key factors that have the greatest contribution to variation and to ascertain those settings or values that result in the least variability.

In developing methods to better understand the influences upon the functionality of products and associated processes, Professor Taguchi has been particularly recognized for three major contributions to the field of quality.

1. Loss function
2. Orthogonal arrays
3. Robustness

In the following, we will discuss Taguchi's contributions to the design of experiments field and illustrate the weaknesses of some of his methods.

5.2 LOSS FUNCTION

Western thinking about product performance relies heavily on specification limits. That is, as long as the product's performance lies within the lower specification and upper specification limits (LSL and USL, respectively), the product is considered good. See Figure 5.1.

Figure 5.1. Zero, One Loss Function

Figure 5.1 illustrates the specification limits concept of accepting a product's performance. Point A is just as good as Point B since both A and B lie within specification limits, although B is much closer to the target and A is much closer to LSL. Point A is much different than Point C. However, logically, since Point B is closer to the target value (T) than A, then Point B is considerably superior to Point A, and there is actually little difference between Point A and Point C. The Taguchi loss function concept overcomes this discrepancy in impact of variability.

Taguchi uses a quadratic loss function of the form

$$L = k(y - T)^2,$$

where k = constant, y = actual measurement, and T = nominal (target) value. This type of loss function will penalize even small departures from the target. See Figure 5.2.

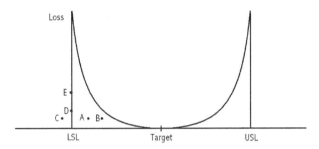

Figure 5.2. Taguchi Loss Function

Figure 5.2 shows that the further the point or measurement is from the target value, the greater the customer dissatisfaction and quality loss. Consequently, the further the point or measurement is from the target, the less desirable that unit is. The loss at Point E in Figure 5.2 is much higher than the one at Point D. In view of this, a key component that engineers did not take into consideration became very important, that is, the reduction of variability as presented in Chapter 4.

We have dedicated two chapters on optimization of the location and minimization of the dispersion. Taguchi, on the other hand, would rely more on a simple parametric function called the signal-to-noise ratio. We will discuss it in subsequent sections, but first, let us present some of the two-level Taguchi designs.

5.3 TAGUCHI DESIGNS

Two-level Taguchi designs have the same objective as fractional factorial designs of resolution III. In fact, all of the two-level Taguchi designs are similar either to 2_{III}^{n-p} designs or Plackett-Burman designs. The only advantage is that they are tabulated providing an easy access to use them.

We will present two tables; the others will be presented in Appendix 3.

Example 5.1: An experimenter needs to study the effect of three factors A, B, and C on the response. A two-level Taguchi design requires four experimental runs. The design is called an L_4 design. Its construction proceeds as follows:

Step 1: Construct a two-level full factorial design with A and B as the basic factors.

Run	A	B
1	+	+
2	−	+
3	+	−
4	−	−

Step 2: Choose $C = -AB$. Note that this is similar to a fractional factorial design of resolution III.

Table 5.1. An L_4 design

Run	A	B	C
1	+	+	−
2	−	+	+
3	+	−	+
4	−	−	−

Example 5.2: An experimenter needs to study the effect of seven factors on the response. A two-level Taguchi design requires eight experimental runs. The design is called an L_8 design. Its construction goes as follows:

Step 1: Construct a two-level full factorial design with A, B, and C as the basic factors.

Run	A	B	C
1	+	+	+
2	−	+	+
3	+	−	+
4	−	−	+
5	+	+	−
6	−	+	−
7	+	−	−
8	−	−	−

Step 2: Choose

$$D = -AB$$
$$E = -AC$$
$$F = -BC$$
$$G = ABC$$

The L_8 design is presented next.

Table 5.2. An L_8 design

Run	A	B	D	C	E	F	G
1	+	+	−	+	−	−	+
2	−	+	+	+	+	−	−
3	+	−	+	+	−	+	−
4	−	−	−	+	+	+	+
5	+	+	−	−	+	+	−
6	−	+	+	−	−	+	+
7	+	−	+	−	+	−	+
8	−	−	−	−	−	−	−

These examples clearly show that the two-level Taguchi designs are not new. In fact, L_4, L_8, L_{16}, L_{32}, etc. are the same as 2_{III}^{n-p} designs except that all even number interactions have been multiplied by -1. Other designs that are not expressed in a power of 2 are similar to Plackett-Burman designs.

The following remarks are very important.

Remark 1: Most of the two-level Taguchi designs are tabulated and hence they are easy to use. See Appendix 3.

Remark 2: Taguchi designs contain very little information on the two-factor interactions. Many statisticians criticize the two-level Taguchi designs. Taguchi claims that in practice, often, interactions do not exist, thus saving an excessive number of experimental runs.

5.4 SIGNAL-TO-NOISE RATIO

Taguchi's advocates believe that a single parametric function contains information about the location and the dispersion parameters. This function is called the signal-to-noise ratio. Signal refers to the location factors, and noise refers to the dispersion factors.

There are three different signal-to-noise ratios that are used in three different situations.

A. Nominal-Is-the-Best

The following signal-to-noise (SN_T) ratio is used when the interest is to set the response to a target value while the variation is minimum:

$$SN_T = 10 \log \left(\frac{\overline{Y}}{S}\right)^2,$$

where $\overline{Y}$ is the average location for an experimental run, S^2 is the variance for the experimental run, and the log is in base 10.

Note that $SN_T = 10 \log \overline{Y}^2 - 10 \log S^2$, if the mean $\overline{Y}$ is set at a target value. Then, maximizing SN_T is equivalent to minimizing $\log S^2$.

Specific examples where nominal-is-the-best should be used are

height	pressure	top-side board temperature
length	density	percent moisture
width	viscosity	voltage
thickness	time	current
diameter	alignment	capacitance
area	frequency	volume
timing	clearance	

Next, we will illustrate the signal-to-noise ratio SN_T by way of an example. We will show that it does not reflect as much information as the methodologies that were presented in the last two chapters. That is, SN_T is not able to determine the location and dispersion factors.

Example 5.3: Consider an experiment with three factors A, B, and C. We will use an L_4 design.

Table 5.3. Average, deviation, and signal-to-noise ratio

Run	A	B	C	Y_1	Y_2	$\overline{\overline{Y}}$	S	SN_T
1	+	+	−	80	90	85	7.07	21.60
2	−	+	+	50	70	60	14.14	12.55
3	+	−	+	75	95	85	14.14	15.58
4	−	−	−	55	65	60	7.07	18.57

To determine which factors affect SN_T, we will evaluate the main effects of A, B, and C.

The effect of A on SN_T is

$$\frac{21.6+15.58}{2} - \frac{12.55+18.57}{2} = 3.03,$$

and similar calculations can be made for B and C.

Table 5.4. Main effects of A, B, and C on SN_T

	A	B	C
effects	3.03	0	− 6.02

Table 5.4 clearly shows that B has no effect on SN_T. Hence B is neither a location factor nor a dispersion factor. This could also be seen by looking at $\overline{Y}$ and S. It is clear that the effect of B on $\overline{Y}$ is 0 and the effect of B on S is 0.

Table 5.4 also shows that factors A and C have an effect on SN_T, but SN_T fails to identify whether A and C are location factors or dispersion factors. However, a simple analysis of $\overline{Y}$ shows that A is a location factor, and a simple analysis of S shows that C is a dispersion factor.

B. Large-Is-the-Best

The following signal-to-noise ratio is used when the interest is to maximize the response while the variation is minimum:

$$SN_L = -10 \log \left(\frac{1}{n} \sum_{i=1}^{n} \frac{1}{y_i^2} \right),$$

where n is the number of replications for a specific run.
Examples of this type of response are

strength	mean time between failures
pull strength	melting point
miles/gallon	corrosion resistance
shelf life	vibration
flash point	ignition temperature

Example 5.4: Let us consider the data in Example 5.3.

Run	A	B	C = −AB	Y_1	Y_2	$\overline{Y}$	S
1	+	+	−	80	90	85	7.07
2	−	+	+	50	70	60	14.14
3	+	−	+	75	95	85	14.14
4	−	−	−	55	65	60	7.07

Next, we will evaluate SN_L for every run. For the first combination $+ + -$, the SN_L value is

$$-10 \log \left[\frac{1}{2} \left(\frac{1}{80^2} + \frac{1}{90^2} \right) \right] = 38.54,$$

and similar calculations can be made for B and C.

Table 5.5. Signal-to-noise SN_L

Run	A	B	C	SN_L
1	+	+	−	38.54
2	−	+	+	35.20
3	+	−	+	38.40
4	−	−	−	35.47

The effect of A on SN_L is

$$\frac{38.54+38.40}{2} - \frac{35.20+35.47}{2} = 3.135,$$

and similarly for B and C:

Table 5.6. Effect of A, B, and C on SN_L

	A	B	C
effects	3.135	− .065	− .205

Table 5.6 clearly indicates that factor A affects SN_L. Thus A is either a location factor or a dispersion factor. However, it fails to show that C is a dispersion factor.

C. Small-Is-the-Best

The following signal-to-noise ratio is used when the objective is to minimize the response while the variation is minimum:

$$SN_S = -10\log\left(\tfrac{1}{n}\sum y_i^2\right),$$

where n denotes the number of replications per run.

A common example is percentage shrinkage. Other examples include

machine wear	product deterioration
residue	access time
percent contamination	response time
lines of computer code	impact damage
loudness	braking

Example 5.5: Let us consider the data in Example 5.3.

Table 5.7. Signal-to-noise SN_S

Run	A	B	C	Y_1	Y_2	SN_S
1	+	+	−	80	90	− 38.60
2	−	+	+	50	70	− 35.68
3	+	−	+	75	95	− 38.64
4	−	−	−	55	65	− 35.59

The effects of A, B, and C on SN_S are presented in Table 5.8.

Table 5.8. Effects of A, B, and C on SN_S

	A	B	C
effects	− 2.98	− .025	− .065

Table 5.8 indicates that only factor A affects SN_S. Thus A is either a location factor or a dispersion factor. Again the signal-to-noise ratio fails to indicate that C is a dispersion factor.

In conclusion, a major disadvantage in using the signal-to-noise ratio is its failure to determine the dispersion factors and the location factors. This is due to the fact that all signal-to-noise ratios that we considered do not distinguish between location factors and dispersion factors.

However, the signal-to-noise ratio can be used to determine the experimental run that minimizes the dispersion. For example, in Example 5.3, using SN_T, the largest value of $SN_T = 21.6$ occurs at (+ + −), and this run produces the smallest deviation S.

In using SN_L in Example 5.4, the largest value of SN_L occurs at (+ + −); this experimental run maximizes the average and minimizes the dispersion S. In using SN_S in Example 5.5, the largest value of SN_S occurs at (− − −); this experimental run minimizes the average and minimizes the dispersion S.

Taguchi supporters often use the average $\overline{Y}$ and the signal-to-noise ratio simultaneously to determine the best settings that optimize the response value while the deviation is minimum. For instance, in Example 5.4, it is clear that A is the only factor that affects the location; thus, setting A at its high level would increase $\overline{Y}$, and similarly only factor A affects SN_L. Thus, setting A at its high level will increase SN_L. The other factors B and C are not location factors. Thus, if it is cheaper to set B at its low level and C at its low level, the value of SN_L may increase. In order to predict the value of SN_L at the combination run (+ − −), a simple prediction equation is needed. We will proceed in a manner similar to that of Chapter 3.

$$\widehat{SN}_L = 36.902 + 1.567A - .0325B - .1025C.$$

Using this equation, the predicted value is 38.10, which is smaller than our experimental champion 38.54. Confirmatory runs for both combinations should be made to determine the best combination.

Next, we will study a major application of Taguchi's approach to robust designs. A product is robust if it is resilient to variations in working environments. Taguchi provides an easy-to-follow mechanism for finding the best settings that both reduce production variation and increase product robustness.

Taguchi's philosophy involves three key ideas:

1. Products should be robust to external variations. For example, commercial transport aircraft fly about as well in a thunderstorm as they do in clean air.
2. Experimental designs are an engineering tool to accomplish this objective.
3. A product on target is more important than conformance to specifications.

More details on the signal-to-noise ratios are given in Peace [2].

5.5 APPLICATIONS OF TAGUCHI'S APPROACH TO ROBUST DESIGNS

Making products and processes robust requires an active study of environmental factors. For example, a portable cassette player should be waterproof and, also, it should work under other adverse weather conditions.

Taguchi recommends the construction of two arrays, the inner array that consists of the controllable factors and an outer array that consists of the noise factors to help design products with robust design.

A control factor is a factor whose value we want and are able to set and maintain. Therefore, any factors within the process or that go into the process and that we can and want to control are classified as control factors. A noise factor is a factor whose value cannot be set and maintained or a factor whose value we do not wish to be set and/or maintained.

In some cases, it is either impractical or impossible to control the factor. For example, relative humidity within the production area can be regulated only if $1 million is spent to encase the operation in a plastic bubble-like controlled environment. This cost is too high. In this case, relative humidity is a noise factor due to the company's preference not to control it.

The following examples illustrate this major contribution of Taguchi.

A. Analysis: Large-Is-the-Best

Example 5.6: Three raw materials are used in varying amounts to form a certain plastic product. Depending on the quantities of each material used, the product will vary in strength, with the goal to make the strongest product possible. It is also known that ambient temperature and humidity can affect the strength of the product.

Let the three materials be denoted A, B, and C; a study is conducted to select the levels of these factors that make the product's manufacture robust against changes in temperature and humidity.

First, an L_4 design is used for the inner array. Similarly, an L_4 design is used for the outer array.

The inner array uses only the controllable factors A, B, and C.

A	B	C
+	+	−
−	+	+
+	−	+
−	−	−

The outer array uses only the noise (uncontrollable) factors.

D	E	*
+	+	−
−	+	+
+	−	+
−	−	−

The experimental design plan requires 16 runs. For example, the combination of levels $+ + -$ will be used with 4 level combinations of D and E, that is

$$(+ + - + +), (+ + - - +), (+ + - + -), (+ + - - -).$$

Note the * column is not used.

Table 5.9. Taguchi design plan

			Outer	Array				
		D	+	−	+	−		
		E	+	+	−	−		
Inner	Array		↓	↓	↓	↓		
A	B	C					$\overline{Y}$	SN_L
+	+	−	→	43.1	91.1	92.6	41.6	67.10	34.70
−	+	+	→	64.1	92.6	83.6	74.6	78.72	37.67
+	−	+	→	74.6	86.6	85.1	82.1	82.10	38.24
−	−	−	→	13.1	82.1	92.6	16.1	50.97	26.04

Here, $SN_L = -10 \log \left(\frac{1}{4} \sum \frac{1}{y_i^2} \right)$.

Notice that the largest SN_L of 38.24 occurs at the combination $(+ \; - \; +)$, which turns out to correspond to the largest $\overline{Y}$ value as well. Thus, it seems that the best setting for A, B, and C to maximize the strength of the plastic and to make the product robust against changes in temperature and humidity is to set A at its high level, B at its low level, and C at its high level.

A wise investigator, before giving any recommendation, should also analyze the effect of the controllable factors A, B, and C on the signal-to-noise ratio SN_L.

The effect of factor A on SN_L is

$$\frac{34.70 + 38.24}{2} - \frac{37.67 + 26.04}{2} = 4.615,$$

and similar calculations should be made for factors B and C. We obtain the table below.

Table 5.10. Effect of A, B and C on SN_L

	A	B	C
Effects	4.615	4.045	7.585

To determine whether A, B, and C have a strong impact on SN_L, we will evaluate their contributions.

To determine the contribution of factor A on SN_L, let

$$T = (4.615)^2 + (4.045)^2 + (7.585)^2 = 95.20.$$

Then, the contribution of factor A on SN_L is

$$\frac{21.30}{95.20} \times 100\% = 22\%,$$

and similar calculations should be made for factors B and C. We obtain the table below.

Table 5.11. Contribution of main effects on SN_L

	A	B	C
Contribution	22%	17%	61%

It is clear that A, B, and C have a strong effect on SN_L, and in order to maximize SN_L, all factors A, B, and C must be set at their high level. Consequently, the experimental run that maximizes SN_L must be $(+ \; + \; +)$, which has not been used in our experimental plan.

To predict the value of SN_L at the experimental run $(+ + +)$, we need to determine the relationship between SN_L and the control factors A, B, and C.

Proceeding as we did in Chapters 3 and 4, the prediction equation is

$$\widehat{SN}_L = 34.16 + 2.08A + 2.02B + 3.79C.$$

Hence, the maximum value of SN_L is

$$\widehat{SN}_L = 34.16 + 2.31 + 2.02 + 3.79 = 42.28.$$

Since the predicted SN_L at the experimental run $(+ + +)$ is larger than that of $(+ - +)$, we would recommend $(+ + +)$ as the best setting for A, B, and C to maximize the strength of the plastic and to make the product robust against changes in temperature and humidity.

B. Analysis: Small-Is-the-Best

Example 5.7: For a specific control device, the time required for the circuit to switch the "On" state to the "Off" state when the L.E.D. (Light Emitting Diode) drive current is removed is critical. The product specification requirement states that the time between changing states can be no more than .005 seconds. However, functional testing of current production reveals a large percentage of the units have not been in compliance with the timing specification.

Objective: Minimize the time required for the device to switch from the "On" state to the "Off" state.

The experimental design and data are shown in Table 5.12. The controllable factors are

A:	Package type	type 1	type 2
B:	Burn in	no	yes
C:	Test delay	no delay	delay
D:	Component-supplier	vendor X	vendor Y

The noise factors are

M:	Moisture	low	high
N:	Drying	normal	heat gun
O:	Soldering	hand solder	wave solder

The team decides to run an L_8 design for the inner array and an L_4 design for the outer array.

Since the objective is to minimize the time required for the device to switch from the "On" state to the "Off" state, then we will use small-is-the-best SN_S:

$$SN_S = -10 \log \frac{1}{n} \sum_{i=1}^{n} y_i^2,$$

where n denotes the number of replications per run. For run #1, SN_S is

$$SN_S = -10 \log \frac{2.245^2 + 6.160^2 + 6.100^2 + 2.255^2}{4} = -13.28,$$

and similar calculations are made for all remaining runs. We obtain the following table.

Table 5.13. Signal-to-noise for every experimental run

Run	A	B	$-A \times B$	C	$-A \times C$	$-B \times C$	$D = A \times B \times C$	SN_S
1	+	+	−	+	−	−	+	− 13.28
2	−	+	+	+	+	−	−	− 15.62
3	+	−	+	+	−	+	−	− 14.77
4	−	−	−	+	+	+	+	− 15.32
5	+	+	−	−	+	+	−	− 13.06
6	−	+	+	−	−	+	+	− 13.79
7	+	−	+	−	+	−	+	− 8.98
8	−	−	−	−	−	−	−	− 16.91

The experimental run that seems to minimize time while the variation due to the main factor is minimal is run #7. That is, $(+ - - +)$.

We need to be careful here. We should determine which factors affect SN_S.

The effect of factor A (package type) on SN_S is

$$\frac{-13.28 - 14.77 - 13.06 - 8.98}{4} - \frac{-15.62 - 15.32 - 13.79 - 16.91}{4} = 2.88,$$

and similar calculations are repeated for B, C, and D; we obtain the following table:

Table 5.14. Main effects on SN_S

	A	B	C	D	$-A \times B$	$-A \times C$	$-B \times C$
Effects	2.88	.057	− 1.56	2.24	1.35	1.44	− .54

To determine which factors among A, B, C, and D have a large impact on SN_S, we will evaluate their contributions.
To evaluate the contribution of factor A, first let

$$T = 2.88^2 + .057^2 + 1.56^2 + 2.24^2 + 1.35^2 + 1.44^2 + .54^2 = 19.94.$$

Then, the contribution of factor A is

$$\frac{8.29}{19.94} \times 100\% = 42\%,$$

and similarly, for factors B, C, and D, we obtain the following table:

Table 5.15. Factor contribution to SN_S

	A	B	C	D	$- A \times B$	$- A \times C$	$- B \times C$
Contribution	42%	0%	12%	25%	9%	10%	2%

Table 5.15 clearly shows that the main effects A, C, and D and the interactions A × B and A × C have a large impact on SN_S. Table 5.14 indicates that A should be set at +1, C should be set at − 1, and D should be set at +1. In order to determine the best setting for factor B, we should analyze carefully the interaction A × B. Table 5.14 indicates that A and B should be set at opposite levels. Thus, either A should be set at +1 and B at − 1 or A should be at − 1 and B at +1. To determine which of the two combinations is the best, let us evaluate

$$A_+B_- = \frac{-14.77 - 8.98}{2} = -11.88,$$

and

$$A_-B_+ = \frac{-15.62 - 13.79}{2} = -14.71.$$

Now, it is clear that A should be set at +1 and B at − 1, which reinforces the previous selection of the setting of factor A when it was based solely on its main effect.

Consequently, (+ − − +) is the best experimental run.

The choice of B at − 1 instead of at + 1 can be also justified through the regression equation:

$$\widehat{SN}_S = -13.96 + 1.44A - .78C + 1.12D - .68AB - .72AC.$$

If B were to be set at + 1, then the predicted value of $\widehat{SN}_S$ at (+ + − +) is

$$\widehat{SN}_S = -13.96 + 1.44 + .78 + 1.12 - .68 + .72 = -10.58,$$

which is smaller than that of the paper champion $(+ - - +)$.

In order to minimize the time required while the variation due to moisture, drying and soldering is minimum, we should employ the following strategy: Use a Type 2 package, do not burn in, do not delay the test, and use vendor Y.

C. Analysis: Nominal-Is-the-Best

Example 5.8: The yardsticks at Three Yarder Yardstick Company have to be 100% inspected. Numerous yardsticks have to be filed down and many others have to be scrapped because they are too short.

Objective: Reduce yardstick length variability and hit consistently a target of 36.00 inches.

The experimental design and data are shown in Table 5.16.

The controllable factors are

A: Blade metal   carbon steel   tungsten steel
B: Saw rpm   low   high
C: Wood kiln time   1 month   3 months
D: Motor horse power   2 hp   3 hp
E: Saw fence angle   0 degrees   12 degrees
F: Operator   Bill   John

The noise factors are

M: Wood source   northern   southern
N: Blade age   new   20,000 units

The team decided to use an L_8 design for the inner array and a full factorial for the outer array.

Since the objective of the study is to hit a target of 36.00 inches consistently with minimum variability, then we will use the nominal-is-the-best option:

$$SN_T = 10 \log \left(\tfrac{\bar{Y}}{S}\right)^2.$$

In performing analysis in the previous examples on the large-is-the-best and the small-is-the-best, significant factors and optimum levels were determined solely by the signal-to-noise ratios. However, for nominal-is-the-best, a different approach is required.

To understand this approach, we will consider a two-stage strategy. Figure 5.3 illustrates this strategy.

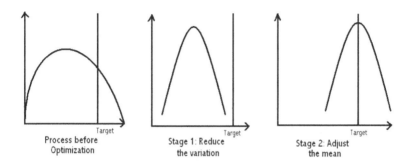

Figure 5.3. Nominal-Is-the-Best Optimization Strategy

First, we need to reduce variability. This is the first step in improving the process or the product performance. By determining which variables under study affect consistency from one unit to the next or uniformly within each product, we have ascertained the knowledge to guide us in reducing variability within the process and ultimately between products. This new insight directs us toward which factors have a significant impact on variability. Besides focusing attention on the appropriate factors, this information can also tell us which factor levels or settings will give us the least variability and will be least affected by noise.

Second, after reducing the variability, we need to adjust the mean. By focusing on those variables that have a significant impact on the average result but little or no impact on the variation, we can move the process closer to the target value.

This new strategy will be used in our example.

Stage 1: Determine the factors that have an impact on the variability, and then choose the settings of these factors to minimize the variability.

1. Calculate SN_T for every run. Since SN_T's calculations involve the average $\bar{Y}$ and the standard deviation S, we first present the average and standard deviation for every run.

Run	A	B	$-AB$	C	D	E	F	$\bar{Y}$	S
1	+	+	−	+	−	−	+	35.540	.413
2	−	+	+	+	+	−	−	35.618	.489
3	+	−	+	+	−	+	−	36.643	.375
4	−	−	−	+	+	+	+	35.994	.512
5	+	+	−	−	+	+	−	36.615	.365
6	−	+	+	−	−	+	+	35.845	.502
7	+	−	+	−	+	−	+	35.565	.738
8	−	−	−	−	−	−	−	35.648	.660

The signal-to-noise ratio for run #1 is

$$10 \log \left(\tfrac{35.540}{.413} \right)^2 = 38.695,$$

and in a similar fashion we obtain the following table for all remaining runs:

Table 5.17. SN_T for every experimental run

Run	A	B	$-AB$	C	D	E	F	SN_T
1	+	+	−	+	−	−	+	38.695
2	−	+	+	+	+	−	−	37.247
3	+	−	+	+	−	+	−	39.799
4	−	−	−	+	+	+	+	36.939
5	+	+	−	−	+	+	−	40.027
6	−	+	+	−	−	+	+	37.074
7	+	−	+	−	+	−	+	33.659
8	−	−	−	−	−	−	−	34.649

To determine which factors have an impact on the variability, we need to calculate the effect of every factor on SN_T.

The effect of factor A on SN_T is

$$\frac{38.695+39.799+40.027+33.659}{4} - \frac{37.247+36.939+37.074+34.649}{4} = 1.57$$

and for factors B, C, D, E, and F, we obtain the table below with similar calculations.

Table 5.18. Main effects on the variability

	A	B	C	D = − AC	E = − BC	F = ABD	− A × B
Effects	1.57	2.0	1.82	.59	2.4	− 1.34	− .63

Next, we will determine the contribution of every factor to the variability. Let

$$T = (1.57)^2 + (2.0)^2 + (1.82)^2 + (.59)^2 + (2.4)^2 + (1.34)^2 + (.63)^2 = 18.08.$$

Then, the contribution of factor A to the variability is

$$\tfrac{2.47}{18.08} \times 100\% = 14\%,$$

and similar calculations will follow for all other factors. We obtain the table below.

Table 5.19. Contribution to the variation

	A	B	C	D	E	F	− A × B
Contribution	14%	22%	18%	2%	32%	10%	2%

Table 5.19 clearly shows that factors A, B, C, and E have a great impact on the variability.

From Table 5.18, it follows that the best settings to minimize the variability should be chosen as follows:

A	should be set at + 1 (second level)
B and C	should be set at opposite levels
E	should be set at + 1 (second level)

To determine which of the combinations is the best, let us evaluate

$$B_+C_- = \tfrac{40.027 + 37.074}{2} = 38.551$$

and

$$B_-C_+ = \tfrac{39.799 + 36.939}{2} = 38.369.$$

Now, it is clear that B should set at + 1 and C at − 1.

Stage 2: Determine the factors that have a great impact on the average (location).

The effect of factor A on the average $\overline{Y}$ is

$$\frac{35.540+36.643+36.615+35.565}{4} - \frac{35.618+35.994+35.845+35.648}{4} = .31$$

and similar calculations follow for the effects of all other factors on the average $\overline{Y}$.

We obtain the table below.

Table 5.20. Main effects on the averages

	A	B	C	D	E	F	$- A \times B$
Effects	.31	$-.06$	.03	.03	.68	$-.40$	$-.032$

Next, we will determine the contribution of every factor on the average $\overline{Y}$. Let

$$T = (.31)^2 + (-.06)^2 + (.03)^2 + (.03)^2 + (.68)^2 + (.40)^2 + (-.032)^2 = .725.$$

Then, the contribution of factor A on $\overline{Y}$ is

$$\frac{.096}{.725} \times 100\% = 13\%,$$

and similar calculations follow for all other factors. We obtain the table below.

Table 5.21. Factor contribution to the average

	A	B	C	D	E	F	$- A \times B$
Contribution	13%	0%	0%	0%	64%	22%	1%

Table 5.21 shows that A, E, and F have a great impact on the average. However, since A and E have already been set at $+1$, F is the sole factor that should adjust the average height.

From Table 5.20, it follows that the prediction equation that relates the average length to the important factors is

$$\widehat{Y} = 35.934 + .155A + .34E - .2F.$$

We want $\widehat{Y}$ to be as close as possible to 36.00 inches, but A and E need to be set at the second level because they minimize the variances. We then obtain

$$36 = 36.429 - .2F.$$

Now, solving for F, we obtain
$$F = 0.2145,$$

which is closer to the second level (i.e., John).

Since factor D has no impact on either the variation or the average, the less expensive level would be chosen. For factor D (motor horsepower), the smaller (cheaper) motor may be the preferred choice.

In conclusion, in order to reduce the yardstick length variability and hit consistently a target of 36.00 inches, we recommend the following settings:

For the blade metal, use a tungsten steel.

For the saw rpm, use high.

For the wood kiln time, use 1 month.

For the motor horsepower, use 2 hp.

For the saw fence angle, use 12 degrees.

And finally, the operator should be John.

5.6 COMMENTS ON THE TAGUCHI METHOD

There has been a good deal of controversy about Taguchi's methodology since its introduction to the United States. The concern centers around Taguchi's use of statistical methods, not his philosophy of "loss function" or robust design.

One concern involves the L_N designs, which are mainly of resolution III, and the combination of inner array and outer array requires a large number of experimental runs.

Another problem is that signal-to-noise ratios may not always work in the desired manner. See, for example, Box [3] and Nair et al. [4].

Apart from its statistical shortcomings, Taguchi's philosophy has had a strong influence on the use of Design of Experiments methods worldwide. Perhaps the most important reason for this success is that he provides a "cookbook" approach to the Design of Experiments. Although the statistical issues are not completely settled, the end result has been a greatly enhanced use and appreciation of Design of Experiments techniques in general.

5.7 PROBLEMS

1. What is Taguchi's contribution to Design of Experiments?
2. What are the disadvantages of using the signal-to-noise ratio?
3. Recall that $SN_T = 10 \log \left(\frac{\bar{Y}}{S} \right)^2$. Why is maximizing SN_T equivalent to minimizing the dispersion?
4. What types of factors are used in the inner array and what types of factors are used in the outer array?
5. What is meant by robust design?
6. Redo problem 1c of Chapter 4 using the appropriate signal-to-noise ratio. Compare the results using the signal-to-noise ratio and the results you found originally.

7. Consider the experimental design plan of problem 1 from Chapter 3. What are the optimal settings to reduce the time variability and hit consistently a target of 60.0 minutes?

REFERENCES

[1] Taguchi, G. and Konishi, S. (1987), *Taguchi Methods: Orthogonal Arrays and Linear Graphs*, America Supplier Institute, Inc., Dearborn, MI.
[2] Peace, G.S. (1993), *Taguchi Methods*, Addison-Wesley, Reading, MA.
[3] Box, G.E.P. (1988), Signal-to-noise ratios, performance criteria, and transformations (with discussion), *Technometrics* **30**, 1-40.
[4] Nair, V.N., et al. (1992), Taguchi parameter design: A panel discussion, *Technometrics* **34**:2, 127-161.

CHAPTER 6

Statistical Optimization of the Location Parameter

6.1 INTRODUCTION

In Chapter 3, we dealt with the problem of deciding whether the main effects of factors and the effect of the two-factor interactions should enter the prediction equation $\widehat{Y}$ so that the best settings of factors that optimize the location parameter can be obtained. However, the decision that was made in Chapter 3 was based on two non-statistical techniques, the Pareto chart and the factor contribution technique.

In this chapter, we provide the experimenter with a widely used statistical technique called the analysis of variance (ANOVA) and its application in location optimization for two-level factorial designs.

We begin with a simple example that illustrates the analysis of variance technique.

Example 6.1: In the manufacture of synthetic fiber, the material that is still in the form of a continuous flow is subjected to high temperatures in order to improve its shrinkage properties. The data below are the results of two temperature tests at 120°C and 140°C, in percent shrinkage of the fiber.

Table 6.1. One-factor ANOVA data

	Temperature	% Shrinkage of Fibers
Run	Factor A	Replicated Response Values
1	120°C (− −)	3.45 3.43 3.53 3.57
2	140°C (+ + +)	3.67 4.03 3.91 3.96

The average percent shrinkage and the dispersions are tabulated below.

Table 6.2. Average shrinkage and dispersion

Run	A	$\overline{Y}$	S
1	−	3.50	.066
2	+	3.90	.156

Table 6.2 shows that the average percent shrinkage depends on the setting of the temperature. As a matter of fact, the effect of the temperature is $\Delta = .4$. Is the difference statistically significant? If we conclude that

significant difference exists, then the temperature is an important predictor of the shrinkage.

The experimenter needs to perform a statistical test called hypothesis testing (HT). In HT, there will be two claims. These are

Ho: The temperature does not affect the mean percent shrinkage (null hypothesis).

Ha: The temperature affects the mean percent shrinkage (alternate hypothesis).

If we conclude that Ha is true, then the temperature affects the mean percent shrinkage. If, on the other hand, we conclude that Ho is true, then the temperature does not affect the mean percent shrinkage.

The decision to choose whether the null hypothesis, Ho, is true or the alternative hypothesis, Ha, is true can be made by following these three steps:

Step 1: Compute To $= \Delta \sqrt{\frac{n}{S_1^2 + S_2^2}}$, where

n is the number of replications per run, $n = 4$, $\Delta = .4$, $S_1^2 = .0044$, and $S_2^2 = .0245$.

In our example,

$$\text{To} = 4.72.$$

Step 2: Determine T^* from Appendix 5, where $T^* = t(\frac{\alpha}{2}, 2n - 2)$ and α is the level of significance, which is usually set at .05 (or 5%).

In our example, $T = t(.025, 6) = 2.447$.

Step 3: Compare $|T_0|$ to T^*.

$$\text{If } |T_0| \leq T^*, \text{ then we accept Ho.}$$
$$\text{If } |T_0| > T^*, \text{ then we accept Ha.}$$

In our example, $|T_0| = 4.72 > 2.447$; hence, we accept Ha. Therefore, the temperature affects the mean percent shrinkage and we are 95% confident about our decision.

Furthermore, the prediction equation that relates the average percentage of shrinkage (location parameter) and the temperature is

$$\hat{Y} = 3.7 + .2A.$$

Another example that illustrates the analysis of variance technique for one-factor study is presented.

Example 6.2: A cigarette manufacturer sent each of two laboratories presumably identical samples of tobacco. Each laboratory made five determinations of the nicotine content in milligrams as follows:

Table 6.3. One-factor ANOVA data

	Nicotine Content
Laboratory I	24 27 26 21 24
Laboratory II	27 28 23 31 26

The manufacturer wishes to determine if the mean of the nicotine content (location parameter) is the same for both laboratories.

The average nicotine contents and the dispersions are tabulated below.

Table 6.4. Average nicotine contents and dispersion

	$\bar{Y}$	S
Laboratory I	24.4	2.30
Laboratory II	27.0	2.91

Table 6.4 shows that the average nicotine contents are different. In fact, the difference $\Delta = 2.6$. Is this difference statistically significant?

We will follow the three steps that were presented above.

Step 1: Compute $To = \Delta \sqrt{\frac{n}{S_1^2 + S_2^2}}$, where $n = 5$, $\Delta = 2.6$, $S_1^2 = 5.3$, and $S_2^2 = 8.5$. Hence, $To = 1.56$.

Step 2: Determine T^* from Appendix 5, where $T^* = t(.025, 8)$. $T^* = 2.306$.

Step 3: Compare $|To|$ to T^*.

Since $|To| = 1.56$ is less than 2.306, we conclude that the mean of the nicotine content is the same for both laboratories.

Next, we present the analysis of variance technique for two-level factorial designs with the objective of optimizing the location parameter.

We shall study three different cases: the replicated two-level full factorial design, the unreplicated two-level full factorial design, and the two-level fractional factorial design.

6.2 REPLICATED TWO-LEVEL FULL FACTORIAL DESIGN

The analysis of variance technique that was described above can be easily extended to the replicated two-level factorial design.

For instance, let us revisit Example 4.1. The average yield and the dispersions are tabulated below.

Table 6.5. Average yield and dispersions

Run	A	B	AB	Y_1	Y_2	Y_3	$\overline{Y}$	S
1	+	+	+	34	30	29	31.00	2.65
2	−	+	−	18	19	23	20.00	2.65
3	+	−	−	36	32	32	33.33	2.30
4	−	−	+	28	25	27	26.66	1.53

In Chapter 3, we discussed a nonstatistical approach for determining which factor effects and/or interactions should go into the prediction equations of the mean yield. There are three main problems in hypothesis testing that need to be addressed.

Test 1: Testing for a significant A effect.

Step 1: Compute To, where $To = \frac{k\Delta_A}{2}\sqrt{\frac{n}{\sum_{i=1}^{k}S_i^2}}$, where k is the number of distinct runs, Δ_A is the main effect of factor A, and n is the number of replications per run.

For our example, $k = 4$, $\Delta_A = 8.83$, and $n = 3$, so

$$To = \frac{4 \times 8.83}{2}\sqrt{\frac{3}{2.65^2 + 2.65^2 + 2.30^2 + 1.53^2}} = 6.57.$$

Step 2: Determine T* from Appendix 5, where $T^* = t(\frac{\alpha}{2}, 2^m(n-1))$ and m is the number of factors. Hence, using $\alpha = .05$ it follows that

$$T^* = t(.025, 8) = 2.306.$$

Step 3: Compare $|To|$ to T*.

Since $|To| = 6.57$ is greater than 2.306, we conclude that the main effect of factor A is significant and hence it belongs to the prediction equation.

Test 2: Testing for a significant B effect. Here also, we will proceed as we did in Test 1.

Step 1: Compute $To = \frac{k\Delta_B}{2}\sqrt{\frac{n}{\sum_{i=1}^{k}S_i^2}}$, where $k = 4$, $\Delta_B = -4.49$, $n = 3$.

Hence $To = -3.34$.

Step 2: Determine T* from Appendix 5.

$$T^* = t(\frac{\alpha}{2}, 2^m(n-1)).$$

Using $\alpha = .05$, it follows that $T^* = t(.025, 8) = 2.306$.

Step 3: Compare $|To|$ to T*.

Since $|\text{To}|$ is greater than 2.306, we conclude that the main effect of factor B is significant and hence it too belongs to the prediction equation of the mean yield.

Test 3: Testing for a significant A × B interaction effect.

Step 1: Compute To, where $\text{To} = \frac{k\Delta_{AB}}{2}\sqrt{\frac{n}{\sum\limits_{i=1}^{k}S_i^2}}$, where $k = 4$, $\Delta_{AB} = 2.16$,

$n = 3$. Hence, $\text{To} = \frac{4\times2.16}{2} \times \sqrt{\frac{3}{2.65^2+2.65^2+2.30^2+1.53^2}} = 1.61$.

Step 2: Determine T* from Appendix 5:

$$\text{T}^* = t(\tfrac{\alpha}{2}, 2^m(n-1)).$$

It follows that $\text{T}^* = t(.025, 8) = 2.306$.

Step 3: Compare To to T*.

Since $|\text{To}| = 1.61$ is less than 2.306, we conclude that the interaction effect is not significant and hence it does not belong in the prediction equation.

In conclusion, both factors A and B are important; however their interaction is not important. Hence, the prediction equation that relates the average yield and the important factors is

$$\widehat{Y} = 27.75 + 4.42A - 2.25B.$$

If the objective is to maximize the average yield (location parameter), then factor A should be set at +1 and factor B should be set at -1.

Another example that illustrates the analysis of variance technique for location optimization when a replicated two-level full factorial is used is presented below.

Example 6.3: A paint manufacturing company tests new formulas for outside paint by painting two panels of each of two kinds of wood and exposing them for 2 years in two climates (warm, cold), putting two panels for each type of wood in each climate. A group of paint technologists then scores, the panels on a scale from 0 to 100. The objective of the study is to analyze the data and to give recommendations.

A two-level full factorial design is performed. The three factors and their levels are

A: Type of wood   Type 1   Type 2
B: Climate   Warm   Cold
C: Formula   Cheap   Expensive

Table 6.6. Average score and dispersions

A	B	C	Y_1	Y_2	$\bar{Y}$	S
+	+	+	21	15	18	4.24
−	+	+	14	18	16	2.83
+	−	+	20	18	19	1.41
−	−	+	21	23	22	1.41
+	+	−	53	59	56	4.24
−	+	−	56	54	55	1.41
+	−	−	61	63	62	1.41
−	−	−	61	59	60	1.41

We want to determine the best combination of wood type, climate and formula that produces the optimal score.

Test 1: Testing for a significant type of wood effect.

Step 1: Compute To $= \frac{k\Delta_A}{2}\sqrt{\frac{n}{\sum\limits_{i=1}^{k} S_i^2}}$, where $k = 8$, $\Delta_A = .5$, $n = 2$. Then To $= .38$.

Step 2: Determine T* from Appendix 5, where $T^* = t(\frac{\alpha}{2}, 2^m(n-1))$. For our example, $T^* = t(.025, 8) = 2.306$.

Step 3: Compare |To| to T*.

Since |To| $= .38$ is less than 2.306, we conclude that the main effect of the type of wood is not significant.

Test 2: Testing for a significant climate effect.

Step 1: Compute To $= \frac{k\Delta_B}{2}\sqrt{\frac{n}{\sum S_i^2}}$; here $\Delta_B = -4.5$, hence To $= -3.47$.

Step 2: $T^* = t(.025, 8) = 2.306$.

Step 3: |To| $= 3.47 > 2.306$, therefore we conclude that the main effect of paint formula is very significant.

Test 3: Testing for a significant formula effect.

Step 1: Compute To $= \frac{k\Delta_C}{2}\sqrt{\frac{n}{\sum S_i^2}}$; here $\Delta_C = -39.5$, hence To $= -30.02$.

Step 2: $T^* = t(.025, 8) = 2.306$.

Step 3: |To| $= 30.02 > 2.306$, therefore we conclude that the main effect of paint formula is very significant.

Three tests are also needed for two-factor interactions. The formula for To remains the same except for the two-factor interaction effects Δ_{AB}, Δ_{AC}, and Δ_{BC}.

To determine Δ_{AB}, Δ_{AC}, and Δ_{BC}, we will proceed as we did in the previous chapters.

Table 6.7. Two-factor interactions

Run	A	B	C	AB	AC	BC	$\bar{Y}$	S
1	+	+	+	+	+	+	18	4.24
2	−	+	+	−	−	+	16	2.83
3	+	−	+	−	+	−	19	1.41
4	−	−	+	+	−	−	22	1.41
5	+	+	−	+	−	−	56	4.24
6	−	+	−	−	+	−	55	1.41
7	+	−	−	−	−	+	62	1.41
8	−	−	−	+	+	+	60	1.41

Test 4: Testing for a significant A × B interaction effect.
Step 1: To $= \frac{k\Delta_{AB}}{2}\sqrt{\frac{n}{\sum\limits_{i=1}^{k}S_i^2}}$; here $\Delta_{AB} = 1$, hence To $= .76$.

Step 2: $T^* = t(.025, 8) = 2.306$.
Step 3: $|\text{To}| = .76 < 2.306$, therefore we conclude that the two-factor interaction A × B is not significant.
Test 5: Testing for a significant A × C interaction effect.
Step 1: To $= \frac{k\Delta_{AC}}{2}\sqrt{\frac{n}{\sum\limits_{i=1}^{k}S_2^2}}$; here $\Delta_{AC} = -1$, hence To $= -.76$.

Step 2: $T^* = t(.025, 8) = 2.306$.
Step 3: $|\text{To}| = .76 < 2.306$, therefore we conclude that the two-factor interaction A × C is not significant.
Test 6: Testing for a significant B × C interaction effect.
Step 1: To $= \frac{k\Delta_{BC}}{2}\sqrt{\frac{n}{\sum\limits_{i=1}^{k}S_i^2}}$; here $\Delta_{BC} = 1$, hence To $= .76$.

Step 2: $T^* = t(.025, 8) = 2.306$.
Step 3: $|\text{To}| = .76 < 2.306$, therefore we conclude that the two-factor interaction B × C is not significant.
In conclusion, the climate and the formula (B and C) are the only important factors that affect the average score.
Hence, the prediction equation that relates the average score and the important factors is

$$\hat{Y} = 38.5 - 2.25B - 19.75C.$$

This relationship shows that Formula 1 is better than Formula 2, and better scores are obtained during warm weather.

6.3 UNREPLICATED TWO-LEVEL FULL FACTORIAL DESIGN

When replications are not feasible in an experimental setting, $To = \frac{k\Delta}{2} \sqrt{\frac{n}{\sum_{i=1}^{k} S_i^2}}$ becomes useless since its denominator vanishes.

This section describes the analysis of variance technique with the objective of optimizing the location parameter when the experimenter can run only an unreplicated two-level full factorial design.

The following example illustrates the methodology.

Example 6.4: In an effort to minimize conversion, four factors were studied by a chemical engineer.

	Factors	−	+
A:	catalyst charge (lb)	10	15
B:	temperature (°C)	220	240
C:	pressure (psi)	50	80
D:	concentration (%)	10	12

The experimenter decided to run the experiment using 16 experimental runs. He obtained the data given below.

Table 6.8. Two-level full factorial experiment

A	B	C	D	Y
+	+	+	+	78
−	+	+	+	85
+	−	+	+	51
−	−	+	+	59
+	+	−	+	83
−	+	−	+	89
+	−	−	+	50

Table 6.8. (cont.)

A	B	C	D	Y
−	−	−	+	61
+	+	+	−	80
−	+	+	−	87
+	−	+	−	61
−	−	+	−	68
+	+	−	−	82
−	+	−	−	90
+	−	−	−	61
−	−	−	−	71

Recall that the content of this book assumes that only the main effects and the two-factor interactions are to be investigated. If we had m factors, then we need to investigate m main effects and $\frac{m(m-1)}{2}$ two-factor interaction effects. In our example, we need to investigate four main effects and six two-factor interaction effects. To do so, we first construct the full factorial design and determine all main effects and all interaction effects.

Table 6.9. Design matrix for a two-level full factorial of four factors

A	B	C	D	AB	AC	AD	BC	BD	CD
+	+	+	+	+	+	+	+	+	+
−	+	+	+	−	−	−	+	+	+
+	−	+	+	−	+	+	−	−	+
−	−	+	+	+	−	−	−	−	+
+	+	−	+	+	−	+	−	+	−
−	+	−	+	−	+	−	−	+	−
+	−	−	+	−	−	+	+	−	−
−	−	−	+	+	+	−	+	−	−
+	+	+	−	+	+	−	+	−	−
−	+	+	−	−	−	+	+	−	−
+	−	+	−	−	+	−	−	+	−
−	−	+	−	+	−	+	−	+	−
+	+	−	−	+	−	−	−	−	+
−	+	−	−	−	+	+	−	−	+
+	−	−	−	−	−	−	+	+	+
−	−	−	−	+	+	+	+	+	+

Table 6.9. (cont.)

ABC	ABD	ACD	BCD	ABCD
+	+	+	+	+
−	−	−	+	−
−	−	+	−	−
+	+	−	−	+
−	−	−	−	−
+	+	+	−	+
+	+	−	+	+
−	−	+	+	−
+	+	−	−	−
−	−	+	−	+
−	−	−	+	+
+	+	+	+	−
−	−	+	+	+
+	+	−	+	−
+	+	+	−	−
−	−	−	−	+

Next, we determine all main effects and the interaction effects.

Table 6.10. Main effects and interaction effects

	A	B	C	D	AB	AC	AD	BC
Δ	− 8	24	− 2.25	− 5.5	1	.75	0	− 1.25

	BD	CD	ABC	ABD	ACD	BCD	ABCD
Δ	4.5	− .25	− .75	.5	− .25	.75	− .25

Finally, we are ready to start testing.

Test 1: Testing for a significant catalyst charge effect.

Step 1: Compute $\Delta_e = \sqrt{\Delta_{ABC}^2 + \Delta_{ABD}^2 + \Delta_{ACD}^2 + \Delta_{BCD}^2 + \Delta_{ABCD}^2}$.
For our example,

$$\Delta_e = \sqrt{(-.75)^2 + (.25)^2(-.5)^2 + (-.75)^2 + (-.25)^2} = \sqrt{1.5} = 1.22.$$

If we had m factors and did not have any prior information about the two-factor interaction, then, using a resolution V design, Δ_e becomes:

$$\Delta_e = \sqrt{\Delta_e^2},$$

where Δ_e^2 is the sum of the square of all three-factor and higher interaction effects.

Step 2: Compute To $= \frac{\Delta_A}{\Delta_e} \sqrt{2^m - 1 - m - \frac{m(m-1)}{2}}$.

Here, $2^m - 1 - m - \frac{m(m-1)}{2}$ is the number of three-factor and higher interactions.

In our example, $m = 4$, $\Delta_A = -8$, and $\Delta_e = 1.22$. Hence, To $= \frac{-8}{1.22} \times \sqrt{5} = -14.66$.

Step 3: Determine T* from Appendix 5, where

$$T^* = t\left(\frac{\alpha}{2}, 2^m - 1 - m - \frac{m(m-1)}{2}\right).$$

In our example, $T^* = t\left(\frac{\alpha}{2}, 5\right)$. Using $\alpha = .05$, $T^* = t(.025, 5) = 2.571$.

Step 4: Compare $|To|$ to T^*. Since $|To| = 14.66$ is greater than 2.571, we conclude that the main effect of catalyst charge is significant.

Test 2: Testing for a significant temperature effect.
Step 1: $\Delta_e = 1.22$ as before.
Step 2: $T_0 = \frac{\Delta_B}{\Delta_e} \sqrt{2^m - 1 - m - \frac{m(m-1)}{2}}$.
For our example, $T_0 = \frac{24}{1.22} \sqrt{5} = 43.98$.
Step 3: $T^* = t(.025, 5) = 2.571$ as before.
Step 4: $|T_0| = 43.98$ is greater than 2.571, therefore we conclude that the main effect of temperature is significant.

Test 3: Testing for a significant pressure effect.
Step 1: $\Delta_e = 1.22$ as before.
Step 2: $T_0 = \frac{\Delta_C}{\Delta_e} \sqrt{2^m - 1 - m - \frac{m(m-1)}{2}}$.
In our example, $T_0 = -4.12$.
Step 3: $T^* = t(.025, 5) = 2.571$ as before.
Step 4: $|T_0| = 4.12$ is greater than 2.571, therefore we conclude that the main effect of pressure is significant.

Test 4: Testing for a significant concentration effect.
Step 1: $\Delta_e = 1.22$ as before.
Step 2: $T_0 = \frac{\Delta_D}{\Delta_e} \sqrt{2^m - 1 - m - \frac{m(m-1)}{2}}$.
In our example, $T_0 = -10.08$.
Step 3: $T^* = t(.025, 5) = 2.571$ as before.
Step 4: $|T_0| = 10.08$ is greater than 2.571, therefore we conclude that the main effect of concentration is significant.

Test 5: Testing for a significant catalyst × temperature effect.
Step 1: $\Delta_e = 1.22$ as before.
Step 2: $T_0 = \frac{\Delta_{AB}}{\Delta_e} \sqrt{2^m - 1 - m - \frac{m(m-1)}{2}}$.
In our example, $T_0 = 1.83$.

Step 3: $T^* = t(.025,5) = 2.571$.

Step 4: $|T_0| = 1.83$ is less than 2.571, therefore we conclude that the interaction catalyst × temperature effect is not significant.

Test 6: Testing for a significant catalyst × pressure effect.

Step 1: $\Delta_e = 1.22$ as before.

Step 2: $T_0 = \frac{\Delta_{AC}}{\Delta_e} \sqrt{2^m - 1 - m - \frac{m(m-1)}{2}}$.

In our example, $T_0 = 1.37$.

Step 3: $T^* = t(.025,5) = 2.571$.

Step 4: $|T_0| = 1.37$ is less than 2.571, therefore we conclude that the interaction catalyst × pressure effect is not significant.

Test 7: Testing for a significant catalyst × concentration effect.

Step 1: $\Delta_e = 1.22$.

Step 2: $T_0 = \frac{\Delta_{AD}}{\Delta_e} \sqrt{2^m - 1 - m - \frac{m(m-1)}{2}}$.

In our example, $T_0 = 0$.

Step 3: $T^* = t(.025,5) = 2.571$.

Step 4: $|T_0| = 0$ and is less than 2.571, therefore we conclude that the interaction catalyst × concentration effect is not significant.

Test 8: Testing for a significant temperature × pressure effect.

Step 1: $\Delta_e = 1.22$.

Step 2: $T_0 = \frac{\Delta_{BC}}{\Delta_e} \sqrt{2^m - 1 - m - \frac{m(m-1)}{2}}$.

$T_0 = -2.29$.

Step 3: $T^* = t(.025,5) = 2.571$.

Step 4: $|T_0| = 2.29$ is less than 2.571, therefore we conclude that the interaction temperature × pressure effect is not significant.

Test 9: Testing for a significant temperature × concentration effect.

Step 1: $\Delta_e = 1.22$.

Step 2: $T_0 = \frac{\Delta_{BD}}{\Delta_e} \sqrt{2^m - 1 - m - \frac{m(m-1)}{2}}$.

$T_0 = 8.25$.

Step 3: $T^* = t(.025,5) = 2.571$.

Step 4: $|T_0| = 8.25$ is greater than 2.571, therefore we conclude that the interaction temperature × concentration effect is significant.

Test 10: Testing for a significant pressure × concentration effect.

Step 1: $\Delta_e = 1.22$.

Step 2: $T_0 = \frac{\Delta_{CD}}{\Delta_e} \sqrt{2^m - 1 - m - \frac{m(m-1)}{2}}$.

$T_0 = -.46$.

Step 3: $T^* = t(.025,5) = 2.571$ as before.

Step 4: $|T_0| = .46$ is less than 2.571, therefore we conclude that the interaction pressure × concentration effect is not significant.

In conclusion, all four main effects and the interaction temperature × concentration have significant effect on the average conversion.

Hence, the prediction equation that relates the average conversion and the important factor is

$$\hat{Y} = 72.25 - 4A + 12B - 1.125C - 2.75D + 2.25B \times D.$$

The best setting to maximize conversion is $(-\ \ +\ \ -\ \ -\ )$, which agrees with the conclusion reached by the eyeball analysis.

6.4 TWO-LEVEL FRACTIONAL FACTORIAL DESIGN

Here, we shall present the methodology through two simple examples. The first example represents the replicated case, and the second represents the unreplicated case.

Example 6.5: A chemist wishes to maximize the yield for a chemical process. Three factors are considered.

A: Temperature   130°C − 150°C
B: Catalyst   1% − 2%
C: pH   6.8 − 6.9

The chemist does not anticipate any two-factor interaction. He decides to run a two-level fractional design of resolution III.

He obtains the following data.

Table 6.11. Design matrix for a replicated two-level fractional factorial design

A	B	C	Y_1	Y_2	Y_3	$\bar{Y}$
+	+	+	75.3	77.1	76.3	76.23
−	+	−	61.2	59.6	60.2	60.33
+	−	−	75.4	73.1	72.5	73.66
−	−	+	66.0	63.3	61.5	63.60

In this example, the experimenter needs to perform three statistical tests.

Test 1: Testing for a significant temperature effect.

Step 1:   Compute $\Delta_e = \sqrt{\sum_{i=1}^{kn} Y_i^2 - kn\left[\bar{Y}^2 + \frac{\Delta_A^2}{4} + \frac{\Delta_B^2}{4} + \frac{\Delta_C^2}{4}\right]}$, where

$\sum_{i=1}^{kn} Y_i^2$ is the sum of squares of all response values, k is the number of distinct runs, and n is the number of replications per experimental run.

In our example,

$$\Delta_e = \sqrt{56788 - 3 \times 4\left[(68.45)^2 + \frac{(12.98)^2}{4} + \frac{(-.35)^2}{4} + \frac{(2.92)^2}{4}\right]}$$

$$= 5.64.$$

If we had m factors and did not have any prior information about the two-factor interactions, then, using a resolution V design, Δ_e becomes

$$\Delta_e = \sqrt{\sum_{i=1}^{kn} Y_i^2 - kn\left[\overline{Y}^2 + \frac{\Delta_I^2}{4}\right]},$$

where Δ_I^2 is the sum of square of all main effects and the two-factor interaction effects.

Step 2: $T_0 = \frac{\Delta_A}{2\Delta_e}\sqrt{nk(nk - 4)}$.

For our example, $T_0 = 11.27$.

Again, if we had m factors and used a design of resolution V, then T_0 becomes

$$T_0 = \frac{\Delta_A}{2\Delta_e}\sqrt{nk\left[nk - 1 - m - \frac{m(m-1)}{2}\right]}.$$

Step 3: Determine T^* from Appendix 5, where

$$T^* = t(\tfrac{\alpha}{2}, nk - 4).$$

Using $\alpha = .05$, $T^* = t(.025, 8) = 2.306$.

If we had m factors and used a design of resolution V, then

$$T^* = t\left(\tfrac{\alpha}{2}, nk - 1 - m - \frac{m(m-1)}{2}\right).$$

Step 4: Compare $|T_0|$ to T^*.

Since $|T_0| = 11.27$ is greater than 2.306, we conclude that the main effect of temperature is significant.

Test 2: Testing for a significant catalyst effect.

Step 1: $\Delta_e = 5.64$ as before.

Step 2: $T_0 = \frac{\Delta_B}{2\Delta_e}\sqrt{nk(nk - 4)}$.

For our example, $T_0 = .30$.

Step 3: $T^* = t(.025, 8) = 2.306$ as before.

Step 4: $|T_0| = .3$ is less than 2.306, therefore the main effect of catalyst is not significant.

Test 3: Testing for a significant pH effect.

Step 1: $\Delta_e = 5.64$ as before.

Step 2: $T_0 = \frac{\Delta_c}{2\Delta_e} \sqrt{nk(nk-4)}$.

For our example, $T_0 = 2.54$.

Step 3: $T^* = t(.025,8) = 2.306$ as before.

Step 4: $|T_0| = 2.54$ is greater than 2.306, therefore we conclude that the main effect of pH is significant.

In conclusion, the temperature and pH have significant effects on the average yield. The prediction equation that relates the average yield to the important factors is

$$\hat{Y} = 68.45 + 6.49A + 1.46C.$$

The best setting to maximize the average yield is

$$(+ \ ? \ + \ ).$$

The catalyst could be set either at $+1$ or -1. However since it has a negative effect, we should set it at -1. Note that this run $(+ \ - \ + \ )$ has not been used.

The next example will illustrate the methodology for a two-level fractional design without replications.

Example 6.6: In an effort to minimize the amount of shrinkage in a molded part, an experimentation team decides to select four factors. The team aims to investigate all factors and only the interaction terms set time $\times$ zone 1.

A resolution IV design is used instead of a more expensive design such as the two-level full factorial design.

A: Set time	5 – 10 sec
B: Zone 1 temperature	170°F – 190°F
C: Zone 2 temperature	170°F – 190°F
D: Preheat temperature	150°F – 160°F

They obtain the following data.

Table 6.12. Design matrix for a unreplicated two-level fractional factorial design

A	B	C	D	Y
+	+	+	+	5.3
-	+	+	−	6.2
+	−	+	−	4.2
−	−	+	+	6.9
+	+	−	−	3.8
−	+	−	+	8.8
+	−	−	+	7.3
−	−	−	−	6.6

In this example, the experimentation team needs to perform five statistical tests.

Test 1: Testing for a significant set time effect.

Step 1: Compute $\Delta_e = \sqrt{\sum_{i=1}^{k} Y_i^2 - k\left[\overline{Y}^2 + \frac{\Delta_A^2}{4} + \frac{\Delta_B^2}{4} + \frac{\Delta_C^2}{4} + \frac{\Delta_D^2}{4} + \frac{\Delta_{AB}^2}{4}\right]}$,

where $\sum_{i=1}^{k} Y_i^2$ is the sum of squares of all response values, and k is the number of distinct runs.

For our example,

$$\Delta_e = \sqrt{320.51 - 8\left[37.67 + .975 + .012 + .237 + .879 + .237\right]}$$

$$= .65.$$

If we had m factors and did not have any prior information about the two-factor interactions, then using a resolution V design, Δ_e becomes

$$\Delta_e = \sqrt{\sum_{i=1}^{k} Y_i^2 - k\left[\overline{Y}^2 + \frac{\Delta_I^2}{4}\right]},$$

where Δ_I^2 is the sum of squares of all main effects and the two-factor interaction effects.

Step 2: Compute $T_0 = \frac{\Delta_A}{2\Delta_e}\sqrt{k(k-6)}$.

For our example, $T_0 = -6.07$.

Again, if we had m factors and used a design of resolution V, then T_0 becomes

$$T_o = \tfrac{\Delta_A}{2\Delta_e} \sqrt{k\left[k - 1 - m - \tfrac{m(m-1)}{2}\right]}.$$

Step 3: Determine T* from Appendix 5, where

$$T^* = t(\tfrac{\alpha}{2}, k - 6).$$

Using $\alpha = .1$, $T^* = t(0.05,2) = 2.92$.
If we had m factors and used a design of resolution V design, then

$$T^* = t\left(\tfrac{\alpha}{2},\ k - 1 - m - \tfrac{m(m-1)}{2}\right).$$

Step 4: Compare $|T_0|$ to T^*.
Since $|T_0| = 6.07$ is greater than 2.92, we conclude that the main effect of set time is significant.

Test 2: Testing for a significant zone 1 temperature effect.
Step 1: Compute $\Delta_e = .65$ as before.
Step 2: $T_0 = \tfrac{\Delta_B}{2\Delta_e}\sqrt{k(k - 6)}$.
For our example, $T_0 = -.69$.
Step 3: $T^* = t(.05,2) = 2.92$.
Step 4: $|T_0| = .69$ is less than 2.92, therefore we conclude that the main effect of zone 1 temperature is not significant.

Test 3: Testing for a significant zone 2 temperature effect.
Step 1: Compute $\Delta_e = .65$ as before.
Step 2: $T_0 = \tfrac{\Delta_C}{2\Delta_e}\sqrt{k(k - 6)}$.
For our example, $T_0 = -2.99$.
Step 3: $T^* = t(.05, 2) = 2.92$.
Step 4: $|T_0| = 2.99$ is greater than 2.92, therefore we conclude that the main effect of zone 2 temperature is significant.

Test 4: Testing for a significant preheat temperature effect.
Step 1: Compute $\Delta_e = .65$ as before.
Step 2: $T_0 = \tfrac{\Delta_D}{2\Delta_e}\sqrt{k(k - 6)}$.
For our example, $T_0 = 5.76$.
Step 3: $T^* = t(.05,2) = 2.92$.
Step 4: $|T_0| = 5.76$ is greater than 2.92, therefore we conclude that the main effect factor preheat temperature is significant.

Test 5: Testing for a significant set time $\times$ zone 1 temperature effect.
Step 1: $\Delta_e = .65$ as before.
Step 2: $T_0 = \tfrac{\Delta_{AB}}{2\Delta_e}\sqrt{k(k - 6)}$.
For our example, $T_0 = -2.99$.
Step 3: $T^* = t(.05,2) = 2.92$.

Step 4: $|T_0| = 2.99$ is greater than 2.92, therefore we conclude that the interaction set time $\times$ zone 1 temperature is significant.

In conclusion, the main effects that are significant are set time, zone 2 temperature, and preheat temperature. The interaction set time $\times$ zone 1 temperature was also found significant.

The prediction equation that relates the amount of shrinkage and the important factors is

$$\hat{Y} = 6.14 - .98A - .48C + .94D - .49A \times B.$$

The best setting to minimize the amount of shrinkage is ($+$? $+$ $-$). However, the coefficient of A $\times$ B is negative, so A $\times$ B should be set at $+1$, and therefore B should be set at $+1$. Finally, the best setting to minimize the amount of shrinkage is ($+$ $+$ $+$ $-$). Note that the experimental run ($+$ $+$ $+$ $-$ $-$) has not been used.

Several books present an excellent overview of ANOVA, such as those by Neter, Wasserman, and Kutner [1], Draper and Smith [2], and Montgomery [3].

The following remarks are very important.

Remark 1: In Section 6.2, a general formulation for Δ_e would be $\Delta_e = \sqrt{\Delta_e^2}$, where Δ_e^2 is the sum of the squares of all interactions that are assumed negligible. Furthermore,

$$To = \frac{\Delta_A}{\Delta_e}\sqrt{d} \quad \text{and} \quad T^* = t(\tfrac{\alpha}{2}, d),$$

where d is the number of all interactions that are assumed negligible.

Remark 2: In Section 6.3, a general formulation for Δ_I^2 would be the sum of squares of all main effects and all two-factor interaction effects that the investigator wishes to study. Furthermore,

$$To = \frac{\Delta_A}{\Delta_e}\sqrt{nk[nk - 1 - T]} \quad \text{and} \quad T^* = t(\tfrac{\alpha}{2}, nk - 1 - T),$$

where T is the number of all main effects and two-factor interactions that the investigation wants to study.

6.5   PROBLEMS

1. A two-level factorial design was constructed to test the effect of three different variables on the ultimate tensile strength of a weld. The three variables and the levels used are shown in the tables together with the test results. The three factors are

	Low Level	High Level
A: Ambient temperature (°F)	0	70
B: Wind velocity (mph)	0	20
C: Bar size (in.)	4/8	11/8

Run	A	B	C	Y_1	Y_2
1	+	+	+	93.7	81.7
2	−	+	+	82.7	74.5
3	+	−	+	99.7	95.5
4	−	−	+	77.7	80.5
5	+	+	−	76.0	98.0
6	−	+	−	69.6	86.0
7	+	−	−	90.6	84.0
8	−	−	−	84.0	91.0

Determine the prediction equation that relates the average tensile strength and the important factors and two-factor interactions.

2. A two-level factorial design was constructed to test the effect of three different variables on the alertness of students in an early morning class. The variables tested, the design matrix, and the responses are shown in the tables.

	Low Level	High Level
A: Hours of sleep	4	8
B: Ounces of coffee	6	12
C: Number of donuts	1	2

Run	A	B	C	Alertness (%)
1	+	+	+	75
2	−	+	+	59
3	+	−	+	62
4	−	−	+	43
5	+	+	−	89
6	−	+	−	68
7	+	−	−	72
8	−	−	−	56

Determine the prediction equation that relates the alertness and the important factors and two-factor interactions.

3. Tests were carried out on a newly designed automobile. Five variables were studied:

	Low Level	High Level
A: Engine size (L)	4.0	4.5
B: Fuel octane	87	93
C: Tire pressure (psi)	22	28
D: Driving speed (mph)	45	55
E: Air conditioning	off	on

The objective of the experiment was to determine the effects of these variables on the fuel economy. The experimental results are shown in the following table.

Run	A	B	C	D	E	Y (mpg)
1	+	+	+	+	+	25
2	−	+	+	+	−	32
3	+	−	+	+	−	21
4	−	−	+	+	+	26
5	+	+	−	+	−	28
6	−	+	−	+	+	27
7	+	−	−	+	+	19
8	−	−	−	+	−	27
9	+	+	+	−	−	29
10	−	+	+	−	+	29
11	+	−	+	−	+	20
12	−	−	+	−	−	29
13	+	+	−	−	+	22
14	−	+	−	−	−	26
15	+	−	−	−	−	21
16	−	−	−	−	+	25

The experimenter knows that the two-factor interactions $A \times C$, $B \times C$, and $C \times E$ are not important. Determine the optimal setting for maximizing the fuel economy.

4. Consider the experimental design plan of Example 3.2.
   a. Can you determine whether all main effects and two-factor interactions are important?
   b. If the answer is no, then use the factor contribution methodology to discard a factor that is not important.

c. Now, after discarding the insignificant factor, can you determine whether the effects of all four factors and the two-factor interactions are significant?

REFERENCES

[1] Neter, J., Wasserman, W. and Kutner, M.H. (1985), *Applied Linear Statistical Models*, 2nd edition, Richard D. Irwin, Inc., Homewood, IL.

[2] Draper, N.R. and Smith, H. (1966), *Applied Regression Analysis*, Wiley, New York.

[3] Montgomery, D.C. (1991), *Design and Analysis of Experiments*, 3rd edition, Wiley, New York.

CHAPTER 7

Statistical Minimization of the Dispersion Parameter

7.1 INTRODUCTION

In Chapter 4, guidelines for dispersion minimization were given. In the replicated study, we used the factor contribution technique to determine the main effects and the two-factor interaction effects that are important. In the unreplicated study, we used the Box-Meyer method to determine the main effects that are important.

In this chapter, we will present two statistical techniques that are very useful for minimizing the response variability, the analysis of variance technique and the normal probability plot.

7.2 REPLICATED STUDY

The analysis of variance technique and the normal probability plot of the estimates of the effects that was developed by Daniel [1] will be presented.

We will present two examples. Both examples will show that in general the significant effects are detected by both methods, although in some cases the normal probability plot of the effects is hard to use.

A. Analysis of Variance Techniques

The analysis of variance technique for a replicated two-level factorial design when the objective is the minimization of the variability will be presented through the following example.

Example 7.1: The yield of chemical process has been studied as a function of three factors, at the following experimental levels:

Factors	Low Level	High Level
A: Temperature (°C)	80	120
B: Pressure (psi)	50	70
C: Reaction time (min)	5	15

A two-level full factorial design was performed. The table provides the results of the tests.

Table 7.1. Design matrix for a replicated full factorial

Run	A	B	C	Y_1	Y_2
1	+	+	+	70.51	74.00
2	−	+	+	45.20	49.53
3	+	−	+	43.58	36.99
4	−	−	+	24.80	15.41
5	+	+	−	51.37	48.49
6	−	+	−	27.51	34.03
7	+	−	−	75.62	77.57
8	−	−	−	61.43	58.58

From prior knowledge, it is known that the interactions AB and AC do not affect the dispersion. What factors affect the dispersion?

First, since replications are present, we evaluate the dispersion for each experimental run.

Table 7.2. Dispersion per experiment run

Run	A	B	C	AB	AC	BC	ABC	S
1	+	+	+	+	+	+	+	2.47
2	−	+	+	−	−	+	−	3.06
3	+	−	+	−	+	−	−	4.66
4	−	−	+	+	−	−	+	6.64
5	+	+	−	+	−	−	−	2.04
6	−	+	−	−	+	−	+	4.61
7	+	−	−	−	−	+	+	1.38
8	−	−	−	+	+	+	−	2.01

Next, we proceed exactly as we did in Section 6.3, except that in this case S is of interest instead of Y.

We must determine all main effects and all interaction effects.

Table 7.3. Main effects and interaction effects

	A	B	C	AB	AC	BC	ABC
Δ	− 1.44	− .63	1.70	− .14	.16	− 2.26	.83

Finally, we will perform four statistical tests.

Test 1: Testing for a significant temperature effect.

Step 1: Compute $\Delta_e = \sqrt{\sum_{i=1}^{k} S_i^2 - k\left[\overline{S}^2 + \frac{\Delta_A^2}{4} + \frac{\Delta_B^2}{4} + \frac{\Delta_C^2}{4} + \frac{\Delta_{BC}^2}{4}\right]}$, where

k is the number of distinct runs, $\overline{S}$ is the average of k dispersions, and $\sum_{i=1}^{k} S_i^2$ is

the sum of k variances.

In our example,

$$\Delta_e = \sqrt{112.63 - 8\left[11.28 + .52 + .1 + .72 + 1.28\right]} = 1.19.$$

If we had m factors and did not have any prior information about the effect of some two-factor interactions on the dispersion, then a resolution V design should be used, and Δ_e becomes

$$\Delta_e = \sqrt{\sum_{i=1}^{k} S_i^2 - k\left[\overline{S}^2 + \frac{\Delta_I^2}{4}\right]},$$

where Δ_I^2 is the sum of squares of all main effects and two-interaction effects.
Step 2: Compute To $= \frac{\Delta_A}{2\Delta_e} \sqrt{k(k-5)}$. For our example, To $= -2.96$.
If we had m factors and used a design of resolution V, then To becomes

$$\text{To} = \frac{\Delta_A}{2\Delta_e} \sqrt{k\left[k - 1 - m - \frac{m(m-1)}{2}\right]}.$$

Step 3: Determine T* from Appendix 4, where

$$\text{T*} = t(\tfrac{\alpha}{2}, k - 5).$$

Using $\alpha = .05$, T* $= t(.025, 3) = 3.182$.
If we had m factors and used a resolution V design without having any prior information on the significance of the two-factor interaction effects,

$$\text{T*} = t(\tfrac{\alpha}{2}, k - 1 - m - \tfrac{m(m-1)}{2}).$$

Step 4: Compare |To| with T*.
Since |To| is less than T*, we conclude that the main effect of temperature is not significant.
Test 2: Testing for a significant pressure effect.
Step 1: $\Delta_e = 1.19$ as before.
Step 2: To $= \frac{\Delta_B}{2\Delta_e} \sqrt{k(k-5)} = -1.29$.
Step 3: T* $= t(.025, 3) = 3.182$.

Step 4: $|To| = 1.29$ is less than T*, so we conclude that the main effect of pressure is not significant.

Test 3: Testing for significant reaction time effect.
Step 1: $\Delta_e = 1.19$ as before.
Step 2: $To = \frac{\Delta_C}{2\Delta_e} \sqrt{k(k-5)} = 3.49$.
Step 3: $T^* = t(.025,3) = 3.182$.
Step 4: $|To| = 3.49$ is greater than T*, so we conclude that the main effect of reaction is significant.

Test 4: Testing for a significant pressure $\times$ reaction time effect.
Step 1: $\Delta_e = 1.19$ as before.
Step 2: $To = \frac{\Delta_{BC}}{2\Delta_e} \sqrt{k(k-5)} = -4.64$.
Step 3: $T^* = t(.025,3) = 3.182$.
Step 4: $|To| = 4.64$ is greater than T*, so we conclude that the interaction pressure $\times$ reaction time is significant.

In conclusion, the reaction time effect and the interaction pressure $\times$ reaction time effect are significant.

The prediction equation that relates the yield variability and the important effects is

$$\hat{S} = 3.36 + .85C - 1.13B \times C,$$

the optimal settings for minimizing the yield variability are:

$$(? \quad - \quad -),$$

where temperature can be set either at 80°C or at 120°C. However, since it has a negative effect on the yield variability, we may set it at 120°C, and hence the optimal setting is $(+ , \quad - \quad -)$ which agrees with the seventh experimental run.

Also, note that the predicted minimum yield variability is

$$\hat{S} = 3.36 + .85x(-1) - 1.13x(-1)x(-1)$$

$$= 1.38,$$

which also agrees with the seventh experimental run.

Another example that illustrates the analysis of variance technique when the objective is to minimize the variability is presented.

Example 7.2: A two-level full factorial experiment was designed to test the effect of three different factors on the variability of the ultimate tensile strength of a weld. The three factors and the levels are shown in the tables together with the test results. This experiment was replicated twice.

Factors	Low Level	High Level
A: Ambient temperature (°F)	0	70
B: Wind velocity (mph)	0	20
C: Bar size (in.)	4/8	11/8

Table 7.4. Design matrix for a replicated full factorial

Run	A	B	C	Y_1	Y_2
1	+	+	+	93.70	81.76
2	−	+	+	82.70	74.50
3	+	−	+	99.70	95.50
4	−	−	+	77.70	80.50
5	+	+	−	76.00	98.00
6	−	+	−	69.60	86.00
7	+	−	−	90.60	84.00
8	−	−	−	84.00	91.32

The dispersion for each experimental run is presented in Table 7.5.

Table 7.5. Dispersion parameter

Run	A	B	C	AB	AC	BC	ABC	S
1	+	+	+	+	+	+	+	4.48
2	−	+	+	−	−	+	−	5.79
3	+	−	+	−	+	−	−	2.97
4	−	−	+	+	−	−	+	1.98
5	+	+	−	+	−	−	−	15.56
6	−	+	−	−	+	−	+	11.59
7	+	−	−	−	−	+	+	4.67
8	−	−	−	+	+	+	−	5.17

Then, we determine the effects.

Table 7.6. Main effects and interaction effects

	A	B	C	AB	AC	BC	ABC
Δ	.79	5.66	− 5.44	.54	− .95	− 2.99	− 1.69

Since the three-factor interaction is assumed to be insignificant and the effects of AB and AC are smaller than that of ABC, we may assume that both two-factor interactions AB and AC are not significant.

Consequently, we need to perform only four statistical tests.

Test 1: Testing for a significant ambient temperature effect.

Step 1: $\Delta_e = \sqrt{\sum_{i=1}^{k} S_i^2 - k\left[\overline{S}^2 + \frac{\Delta_A^2}{4} + \frac{\Delta_B^2}{4} + \frac{\Delta_C^2}{4} + \frac{\Delta_{BC}^2}{4}\right]}$.

In our example, $\Delta_e =$

$$\sqrt{491.3 - 8\left[42.60 + .15 + 8.00 + 7.40 + 2.23\right]} = 2.87.$$

Step 2: $\text{To} = \frac{\Delta_A}{2\Delta_e}\sqrt{k(k-5)}$.

In our example, $\text{To} = .67$.

Step 3: $T^* = t(\frac{\alpha}{2}, k - 5)$.

Using $\alpha = .05$, $T^* = t(.025, 3) = 3.182$.

Step 4: Since $|\text{To}| = .67$ is less than T^*, we conclude that the main effect of ambient temperature is not significant.

Test 2: Testing for a significant wind velocity effect.

Step 1: $\Delta_e = 2.87$ as before.

Step 2: $\text{To} = \frac{\Delta_B}{2\Delta_e}\sqrt{k(k-5)} = 4.83$.

Step 3: $T^* = t(.025, 3) = 3.182$.

Step 4: Since $|\text{To}| = 4.83$ is greater than T^*, we conclude that the main effect of wind velocity is significant.

Test 3: Testing for a significant bar size effect.

Step 1: $\Delta_e = 2.87$ as before.

Step 2: $\text{To} = \frac{\Delta_C}{2\Delta_e}\sqrt{k(k-5)} = -4.64$.

Step 3: $T^* = t(.025, 3) = 3.182$.

Step 4: Since $|\text{To}| = 4.64$ is greater than T^*, the effect of bar size is significant.

Test 4: Testing for a significant wind velocity × bar size effect.

Step 1: $\Delta_e = 2.87$ as before.

Step 2: $\text{To} = \frac{\Delta_{BC}}{2\Delta_e}\sqrt{k(k-5)} = -2.55$.

Step 3: $T^* = t(.025, 3) = 3.182$.

Step 4: Since $|\text{To}|$ is less than T^*, we conclude that the interaction effect of BC is not significant.

In conclusion, the wind velocity effect and the bar size effect are significant.

The prediction equation that relates the variability of the tensile strength and the important effects is

$$\hat{S} = 6.52 + 2.83B - 2.72C.$$

B. Normal Probability Plot of Effects

An additional approach for determining the statistical significance of the effects when the characteristic of interest is the dispersion is the normal probability plot of the effects. The normal probability plot can also be used when the characteristic of interest is the location and the design is not replicated.

To compute a normal probability plot, there are eight steps.

Step 1: Calculate the main effects and the interaction effects.

Step 2: Order the effects from the smallest to the largest.

Step 3: Assign a rank i to the ith smallest effect.

Step 4: Evaluate $\frac{i-.5}{p}$, where p is the number of effects.

Step 5: From Appendix 4, determine Z_i, when Z_i is the Z value for which the area under the standard normal curve below Z is $\frac{i-.5}{p}$.

Step 6: Plot Z_i versus the ith smallest effect.

Step 7: Draw a straight line through the majority of points.

Step 8: Points that fall well off the line would suggest the existence of real effects.

To illustrate the eight steps, let us reconsider Example 7.2.

Step 1: The main effects and the interaction effects are

	A	B	C	AB	AC	BC	ABC
Δ	.79	5.66	-5.44	.54	$-.95$	-2.99	-1.69

Step 2-6 are tabulated below.

Table 7.7. Ordered effects and percentiles

	$\Delta_{(i)}$	i	$\frac{i-.5}{7}$	Z_i
C	-5.44	1	.07	-1.48
BC	-2.99	2	.21	$-.81$
ABC	-1.69	3	.36	$-.36$
AC	$-.95$	4	.50	0
AB	.54	5	.64	.36
A	.79	6	.79	.81
B	5.66	7	.93	1.48

Step 7 is tabulated in Figure 7.1.

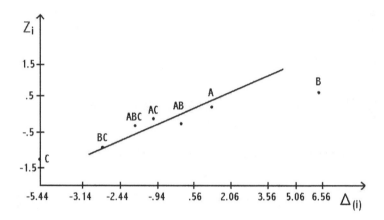

Figure 7.1. Normal Probability Plot of Effects

Step 8: The normal probability plot of the effects suggests that only factors B and C have significant effects. This agrees with the analysis of variance technique.

7.3 UNREPLICATED STUDY

The design methodology for dispersion minimization that was presented in Example 4.3 is a statistical testing that was suggested by Box and Meyer [2]. In fact, Step 8.4 compares

$$Z_0 = \left| \text{Ln} \frac{S_+^2}{S_-^2} \right|$$

with $Z^* = z(1 - \alpha/2)$, where Z^* is determined from Appendix 4, and α is the level of significance.

For instance, if α is set at .05, then

$$Z^* = z(.975) = 1.96.$$

If α is set at .1, then

$$Z^* = z(.95) = 1.645.$$

In Step 8.4, we used 1.96 as the threshold value without mentioning that it is a Z value when the level of significance α is equal to .05.

In the sequel, we will use the normal probability plot to determine which factors contribute to changes in the response variability and then select the factor setting that minimize response variability.

Let us reconsider Example 4.4.

A: Set time   5 − 10 sec

B: Zone 1 temperature   170°F − 190°F

C: Zone 2 temperature   170°F − 190°F

D: Preheat temperature   150°F − 160°F

The response Y is the amount of shrinkage. The design plan is

Run	A	B	C	D	Y
1	+	+	+	+	5.3
2	−	+	+	−	6.2
3	+	−	+	−	4.2
4	−	−	+	+	6.9
5	+	+	−	−	3.8
6	−	+	−	+	8.8
7	+	−	−	+	7.3
8	−	−	−	−	6.6

In Step 7, we calculated the residuals

$$e = Y - \widehat{Y},$$

where $\widehat{Y} = 6.137 - .987A - .487C + .937D - .4875A \times B$.

The residuals were presented in Table 4.6.

Run	A	B	C	D	e
1	+	+	+	+	.19
2	−	+	+	−	.01
3	+	−	+	−	− .01
4	−	−	+	+	− .18
5	+	+	−	−	− .41
6	−	+	−	+	− .23
7	+	−	−	+	.24
8	−	−	−	−	.41

The table of variances of every factor and two-factor interactions is

Table 7.8. Variances per factor level

Variance	A	B	C	D	AC	BC	CD
Low	.08	.07	.15	.11	.07	.02	.03
High	.09	.07	.02	.06	.07	.02	.13

Next, we will follow the eight necessary steps to construct a normal probability plot in order to determine the factors that contribute to changes in the response variability.

Step 1: The natural logarithms of the ratios of the variances are presented.

	A	B	C	D	AC	BC	CD
$\ln\frac{S_+^2}{S_-^2}$	.12	0	-2.01	$-.61$	0	0	1.46

Steps 2-7 are tabulated below.

Table 7.9. Ordered effects and percentiles

	$\delta_{(i)}$	i	$\frac{i-.5}{7}$	Z_i
C	-2.01	1	.07	-1.48
D	$-.61$	2	.21	$-.81$
B	0	4	.5	0
AC	0	4	.5	0
BC	0	4	.5	0
A	.12	6	.79	.81
CD	1.46	7	.93	1.48

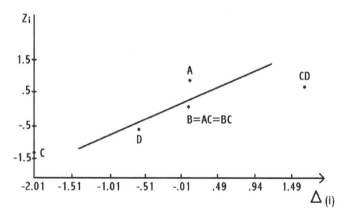

Figure 7.2. Normal Probability Plot of Effects

Step 8: The normal probability plot suggests that factor C affects the variability of the response. It also suggests that the interaction $C \times D$ also affects the variability of the response.

The normal probability plot seems to indicate that the experimenter should set zone 2 temperature at 190°F and the preheat temperature at 150°F.

This approach and Box-Meyer's approach both agree that zone 2 temperature should be set at 190°F; however, the Box-Meyer's approach does not suggest that the interaction zone 2 temperature $\times$ preheat temperature is an important interaction.

A careful experimenter should then perform a further statistical analysis to determine whether the interaction $C \times D$ is significant.

To do so, using the fact that both factors A and B are not important, the design plan can be used as a replicated full factorial design where C and D are the factors. That is, the design plan becomes

Table 7.10. Replicated full factorial design

	C	D	Y_1	Y_2	$\bar{Y}$	S
(1)	+	+	5.3	6.9	6.1	1.13
(2)	−	+	8.8	7.3	8.05	1.06
(3)	+	−	6.2	4.2	5.2	1.41
(4)	−	−	3.8	6.6	5.2	1.98

We first determine the effect of factors C and D and the effect of the interaction $C \times D$.

	C	D	$C \times D$
Δ	− .25	− .6	.32

Now, it is clear that the effect of factor D and the effect of the interaction $C \times D$ are both greater than the effect of factor C. Hence the choice of the optimal design must be based on the terms that are the most important. D is the most important factor; thus it should be set at $+ 1$. CD is the second most important term; thus, it should be set at $- 1$. Hence, factor C should be set at $- 1$. Therefore, the optimal design choice is

$$(?\ ?\ -\ +\ ),$$

where factors A and B can be set either at the high level or at the low level.

The prediction equation that relates the variability of the amount of shrinkage and the important factors is

$$\hat{S} = 1.395 - .125C - .3D + .16C \times D.$$

Note that the predicted minimum variability of the amount of shrinkage is:

$$\hat{S} = 1.395 - .125 \times (-1) - .3 \times (+1) + .16 \times (-1) \times (+1) = 1.06,$$

which also agrees with the second run in Table 7.11.

It should also be noted that our conclusion about the optimal design is contradictory to the choice that we determined in Chapter 4. This contradiction is due to the fact that in Chapter 4 the two factor interactions were not examined in the unreplicated study.

7.4 PROBLEMS

1. Consider the experimental design plan of problem 1 from Chapter 6 with the objective of minimizing the variability of the tensile strength.
   a. Use the analysis of variance technique to determine the main effects and the two-factor interaction effects that are statistically significant (use $\alpha = .1$).
   b. Determine the optimal settings to minimize the variability of the ultimate tensile strength.
   c. Determine the predicted minimum variability of the ultimate tensile strength through the prediction equation that relates the variability of the ultimate strength and the significant main effects and two-factor interaction effects.
   d. Use the normal probability plot of the effects to determine the answers for problems a through c cited above.
2. Consider the experimental design plan of problem 2 of Chapter 6 with the objective of minimizing the variability of students' alertness.
   a. Use the Box-Meyer method to determine the real main effects and the real two-factor interaction effects.
   b. Determine the optimal settings to minimize the variability of the alertness.
   c. Determine the predicted minimum variability of the alertness.
   d. Use the normal probability plot of the effects to determine the answers for problems a through c cited above.

REFERENCES

[1] Daniel, C. (1959), Use of half-normal plots in interpreting factorial two-level experiments, *Technometrics* **26**, 209-216.
[2] Box, G.E.P. and Meyer, R.D. (1986), Dispersion effects from fractional designs, *Technometrics* **28**, 19-27.

CHAPTER 8

Validity of the Prediction Equation

8.1 INTRODUCTION

In previous chapters, optimal settings were obtained through the prediction (regression) equation, which was assumed to provide a good fit. In this chapter, we present simple graphic methods and formal statistical tests for studying the aptness of the prediction equation.

The graphic methods include the plot of the residuals against the fitted values, the normal probability plot of the residuals, and the histogram of the residuals. The formal statistical methods include the adjusted coefficient of determination and the F test for lack of fit.

8.2 GRAPHIC ANALYSIS

A residual e_i is defined as the difference between the observed value and the fitted value:

$$e_i = Y_i - \widehat{Y}_i.$$

There are four important types of departures from an appropriate regression equation:
1. The regression equation is not linear.
2. The error terms do not have a constant variance.
3. The error terms are not independent.
4. The error terms are not normally distributed.

The first three important types of departure can be verified through a simple plot, that is, a plot of the residuals versus the fitted values.

There are four prototypes of residual plots.

Figure 8.1 shows that the residuals tend to fall within a horizontal band centered around zero. This fact alone shows that the predicted values are close to the observed values and, since the residuals fall between two parallel lines, the error terms have a constant variance, and finally, since the residuals seem to randomly fluctuate around zero, the error terms are independent.

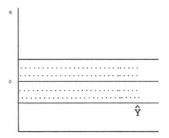

Figure 8.1. Prototype of Residuals Plots (a)

Figure 8.2 shows a prototype situation of a departure from a constancy of the error variances. The error variance increases with $\hat{Y}$.

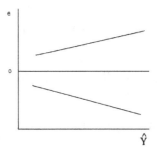

Figure 8.2. Prototype of Residuals Plots (b)

Figure 8.3 shows a prototype situation of a departure from the linearity of the regression equation.

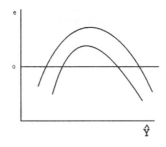

Figure 8.3. Prototype of Residuals Plots (c)

Whenever data are obtained in a time sequence, we should plot the residuals against time. Figure 8.4 shows a lack of independence. There is a clear pattern that shows that at the beginning the residuals are negative and then later they are positive.

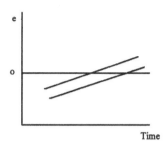

Figure 8.4. Plot of Residuals against Time

Although small departures from normality do not create any serious problems, major problems, on the other hand, should be of concern. The normality of the error terms can be studied formally by examining the normal probability of the residuals.

Let e_1, e_2, ..., e_n denote the residuals for n experimental runs. Construct the normal probability as in Chapter 7.

A plot that is nearly linear suggests agreement with normality.

Figure 8.5 suggests agreement with normality.

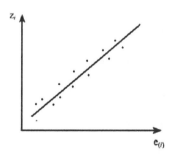

Figure 8.5. Normal Probability of the Residuals

There are other informal approaches that can also investigate departures from normality.

One approach is to determine whether, say, 68% of the residuals fall between $-E$ and E, where

$$E = \sqrt{\frac{\sum\limits_{i=1}^{n} e_i^2}{n-1}},$$

or about 90% of the residuals fall between $-1.64E$ and $1.64E$. This latter approach requires a large number of experimental runs.

The other approach is to construct a histogram of the residuals. If the corresponding histogram is approximately bell-shaped, then the normality assumption of the error terms may be acceptable.

The construction of a histogram of the residuals requires the following steps.

1. Determine the largest residual and the smallest.
2. Define $R = e_{(n)} - e_{(1)}$, where $e_{(n)}$ is the largest residual and $e_{(1)}$ is the smallest.
3. Divide R into a number of intervals that have the same width.
4. Count the number of residuals that fall into the same interval (frequency).
5. Construct a sequence of boxes, where the base of the box is the interval and the height of the box is its corresponding frequency.

The methods that were described will be illustrated through the following example.

Example 8.1: Let us reconsider the data from Example 3.2, where the objective of the study is to maximize the yield of a chemical process. The factors and their levels are displayed below.

A:	Fixed rate	$10 - 15$ liters/min
B:	Catalyst	$1\% - 2\%$
C:	Agitation	$100 - 120$ rpm
D:	Temperature	$140° - 180°C$
E:	Concentration	$3\% - 6\%$

A two-level fractional factorial design of resolution V was chosen for the experiment with one generator

$$E = A\,B\,C\,D.$$

The 16 experimental runs and their corresponding yields are presented next.

Run	A	B	C	D	E	Y
1	+	+	+	+	+	82
2	−	+	+	+	−	95
3	+	−	+	+	−	60
4	−	−	+	+	+	49
5	+	+	−	+	−	93
6	−	+	−	+	+	78
7	+	−	−	+	+	45
8	−	−	−	+	−	69
9	+	+	+	−	−	61
10	−	+	+	−	+	67
11	+	−	+	−	+	55
12	−	−	+	−	−	53
13	+	+	−	−	+	65
14	−	+	−	−	−	63
15	+	−	−	−	−	53
16	−	−	−	−	+	56

In Chapter 3, we found that the catalyst, the temperature, and the concentration are the only important factors. Furthermore, we determined the prediction equation that relates the yield to the important factors

$$\hat{Y} = 65.25 + 10.25B + 6.125D - 3.125E + 5.375BD - 4.75DE.$$

The maximum yield is obtained by setting all three factors, namely the catalyst, the temperature, and the concentration, to the high level (+). The predicted maximum yield is

$$\hat{Y} = 65.25 + 10.25 \times (+ 1) + 6.125 \times (+ 1) - 3.125$$

$$\times (- 1) + 5.375 \times (+ 1) \times (+ 1) - 4.75 \times (+ 1) \times (- 1) = 95.$$

We need to check the validity of the regression equation. To do so, first we plot the residuals versus the fitted values. The plot shows that the assumptions of linearity of the regression constancy of the error variances and independence of the residuals are indeed valid. See Figure 8.6.

Run	Y	$\widehat{Y}$	e
1	82	79.12	2.88
2	95	94.87	.13
3	60	63.62	− 3.62
4	49	47.87	1.13
5	93	94.87	− 1.87
6	78	79.12	− 1.12
7	45	47.87	− 2.87
8	69	63.62	5.38
9	61	62.37	− 1.37
10	67	65.62	1.38
11	55	55.87	− .87
12	53	52.62	.38
13	65	65.62	− .62
14	63	62.37	.63
15	53	52.62	.38
16	56	55.87	.13

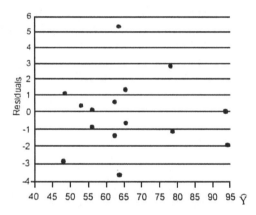

Figure 8.6. Plot of the Residuals against the Fitted Values

The only assumption that is left to verify is the normality of the error terms.

We will construct the normal probability plot of the residuals. To do so, we will follow the six steps that are described next.

Step 1: Order the residuals from the smallest to the largest.

Step 2: Assign a rank i to the ith smallest residual.

Step 3: Evaluate $\frac{i-.5}{n}$, where n is the number of runs.

Step 4: From Appendix 4, determine Z_i.

Step 5: Plot Z_i against the ith smallest residual.
Step 6: If the points fall reasonably close to a straight line, then this will suggest that the error terms are approximately normally distributed.
Steps 1-4 are presented in Table 8.1.

Table 8.1. Ordered residuals and percentiles

e_i	i	$\frac{i-.5}{16}$	Z_i
-3.62	1	.03125	-1.86
-2.87	2	.09375	-1.32
-1.87	3	.15625	-1.01
-1.37	4	.21875	$-.78$
-1.12	5	.28125	$-.58$
$-.87$	6	.34375	$-.4$
$-.62$	7	.40625	.24
.13	8.5	.5	0
.13	8.5	.5	0
.38	10.5	.625	.32
.38	10.5	.625	.32
.63	12	.71875	.58
1.13	13	.78125	.78
1.38	14	.84375	1.01
2.88	15	.90625	1.32
5.38	16	.96875	1.86

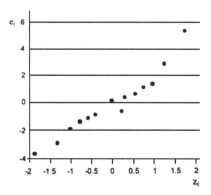

Figure 8.7. Normal Probability of the Residuals

Figure 8.7 shows that the points are reasonably close to a straight line. This suggests that the error terms are approximately normally distributed.

As it was mentioned earlier, there are other informal approaches that can also be used to investigate strong departures from normality.

The first informal approach is to calculate

$$E = \sqrt{\frac{\sum\limits_{}^{16} e_i^2}{15}} = 2.16,$$

and then determine the proportion of the residuals that fall between $-1.64E$ and $1.64E$, that is, between -3.54 and 3.54. There are $\frac{14}{16} \times 100\% = 88\%$ which is close to 90%.

The other informal approach is to construct the histogram of the residuals. Table 8.2 presents the intervals and their corresponding frequencies.

Table 8.2. Interval frequencies

Intervals	Frequency
less than -3	1
-3 to -1	4
-1 to 1	7
1 to 3	3
more than 3	1

The histogram presented in Figure 8.8 shows that the distribution of the residuals is close to a bell – shaped distribution.

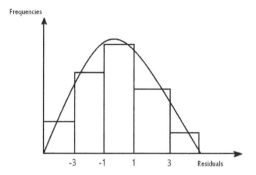

Figure 8.8. Histogram of the Residuals

The plot of the residuals versus the fitted values is often used to check that the error terms have a constant variance and that they are independent.

The assumption of linearity is often verified by calculating the adjusted coefficient of determination, which will be presented next.

8.3 ADJUSTED COEFFICIENT OF DETERMINATION

The adjusted coefficient of determination, denoted by R_a^2, is defined as follows:

$$R_a^2 = 1 - \left(\frac{n-1}{n-p}\right)\frac{\text{SSE}}{\text{SST}},$$

where n denotes the number of experimental runs, p denotes the number of terms in the prediction equation including the constant term, SST denotes the corrected total sum of squares,

$$\text{SST} = \sum_{i=1}^{n} Y_i^2 - n\overline{Y}^2,$$

where $\overline{Y}$ is the average of all response values, and SSE denotes the error sum of squares,

$$\text{SSE} = \sum_{i=1}^{n} e_i^2.$$

R_a^2 takes on the value 1 when all observations fall directly on the prediction equation, i.e., when $Y_i = \widehat{Y}_i$ for all experimental runs.

An R_a^2 value that is greater than .8 is often used as a good indication that the prediction equation provides an adequate fit for the data.

Example 8.2: Let us reconsider the data from Example 8.1. The regression equation was given by

$$\hat{Y} = 65.25 + 10.25B + 6.125D - 3.125E + 5.375BD - 4.75DE.$$

We will evaluate the strength of this relationship by calculating the adjusted coefficient of determination. Here $n = 16$, $p = 6$,

$$\text{SSE} = e_1^2 + e_2^2 \ldots + e_{16}^2 = 70.25, \text{ and}$$
$$\text{SST} = Y_1^2 + Y_2^2 + \ldots + Y_{16}^2 - 16\overline{Y}^2 = 3331.$$

Hence $R_a^2 = 1 - \left(\frac{16-1}{16-6}\right)\frac{70.25}{3331} = 1 - .03 = .97 = 97\%$.

Consequently, the proportion of variability in the yield which is explained by the regression equation is 97%. This clearly shows that the model fits the data points well.

However, the calculation of the adjusted coefficient of determination should also be supplemented with a formal test for determining whether or not a specified prediction equation adequately fits the data. The formal test is called the F test for lack of fit.

8.4 *F* TEST FOR LACK OF FIT

The F test for lack of fit assumes that the error terms are independent, normally distributed, and have a constant variance. Hence this test must be performed only if the experimenter checked the validity of steps 2 through 4.

We illustrate this test for ascertaining whether a linear regression is a good fit for the data. The lack of fit test requires repeated experimental runs. Before embarking on testing whether there is a lack of fit, we need to introduce additional statistical quantities. These are the pure error sum of squares and the lack of fit sum of squares.

Let n denote the total number of experimental runs, let d denote the number of distinct experimental runs, and let n_i denote the number of replications for the ith distinct run, where $i = 1, 2 \ldots, d$.

The pure error sum of squares is denoted by

$$\text{SSPE} = \sum_{i=1}^{d} \sum_{j=1}^{n_i} (Y_{ij} - \overline{Y}_i)^2,$$

where Y_{ij} and $\overline{Y}_i$ can be explained as follows:

Number of distinct runs	Repeated observations	Average
1st	$Y_{11}, Y_{12}, \ldots, Y_{1n_i}$	$\bar{Y}_1$
2nd	$Y_{21}, Y_{22}, \ldots, Y_{2n_i}$	$\bar{Y}_1$
⋮		⋮
ith	$Y_{i1}, Y_{i2}, \ldots, Y_{in_i}$	$\bar{Y}_i$
dth	$Y_{d1}, Y_{d2}, \ldots, Y_{dn_d}$	$\bar{Y}_d$

The lack of fit sum of squares is denoted by

$$SSLF = SSE - SSPE.$$

The F test for lack of fit can be performed by calculating F_0, where

$$F_0 = \frac{(SSLF)/(d-p)}{(SSPE)/(n-d)}.$$

A small F_0 value shows that there is no lack of fit, whereas a larger F_0 value shows that there is a strong lack of fit. Hence, a threshold value to which F_0 should be compared must be determined.

The F test for lack of fit can be performed as follows:

Step 1: Compute F_0.

Step 2: Determine F^* from Appendix 6, where

$$F^* = F(1 - \alpha, d - p, n - d),$$

and α is the level of significance that is usually set to .05.

Step 3: Compare F_0 to F^*.

If $F_0 > F^*$, then there is lack of fit; otherwise, there is no evidence of lack of fit.

Example 8.3: Let us reconsider the data from Example 8.1. We wish to test the following hypothesis:

Ho: There is no evidence of lack of fit.

Ha: There is evidence of lack of fit.

Since the lack of fit test requires repeated experimental runs, let us determine whether our design is a replicated design.

There are three important factors, B, D, and E; hence, we should discard the benign factors A and C. The resulting experimental design plan is presented in Table 8.3.

Table 8.3. Projected experimental design plan

Run	B	D	E	Y
1	+	+	+	82
2	+	+	−	95
3	−	+	−	60
4	−	+	+	49
5	+	+	−	93
6	+	+	+	78
7	−	+	+	45
8	−	+	−	69
9	+	−	−	61
10	+	−	+	67
11	−	−	+	55
12	−	−	−	53
13	+	−	+	65
14	+	−	−	63
15	−	−	−	53
16	−	−	+	56

By rearranging the order of the experimental runs, the design plan becomes a replicated two-level full factorial design. See Table 8.4.

Table 8.4. Replicated two-level full factorial design

B	D	E	Y_1	Y_2	$\bar{Y}$
+	+	+	82	78	80
−	+	+	49	45	47
+	−	+	67	65	66
−	−	+	55	56	55.5
+	+	−	95	93	94
−	+	−	60	69	64.5
+	−	−	61	63	62
−	−	−	53	53	53

Here, $n = 16$, $d = 8$, $n_1 = n_2 = \ldots = n_8 = 2$, and $p = 6$.
$$\text{SSPE} = (82 - 80)^2 + (78 - 80)^2 + (49 - 47)^2 + (45 - 47)^2 + \ldots + (53 - 53)^2 = 63 \text{ and}$$

$$\begin{aligned} \text{SSLF} &= \text{SSE} - \text{SSPE} \\ &= 70.25 - 63 = 7.25. \end{aligned}$$

The F test for lack of fit can now be performed.

Step 1: $F_0 = \frac{(SSLF)/(d-p)}{(SSPE)/(n-d)} = \frac{(7.25)/(8-6)}{(63)/(16-8)} = .46.$

Step 2: $F^* = F(.95, 8 - 6, 16 - 8) = F(.95, 2, 8) = 4.46.$

Step 3: Since $F_0 < F^*$, we conclude Ho, i.e., there is no evidence of lack of fit.

In conclusion, the equation

$$\hat{Y} = 65.25 + 10.25B + 6.125D - 3.125E + 5.375BD - 4.75DE$$

is a very good prediction equation.

Next, we will present recommendations to experimenters in order to check the validity of the requisition equation.

8.5 ANALYSIS RECOMMENDATION

Once a prediction equation is obtained, the experimenters should check the validity of the prediction equation by using the following steps.

Step 1: Plot the residuals versus the fitted values. If the residuals fall between two horizontal lines in a random fashion, then constancy of the error variances and independence are both satisfied. See Figure 8.1. If, on the other hand, the residuals show a clear pattern similar to that of Figures 8.2 and 8.3, then the experimenter should define a new response Y^*, where $Y^* = \sqrt{Y}$, or $Y^* = \log Y$, or $Y^* = \frac{1}{Y}$ and then determine a new prediction equation using Y^* as the response instead of the original response Y.

Once the new prediction equation is obtained, the experimenter should calculate the new residuals and then plot them versus the new fitted values. If the new plot is similar to Figure 8.1, then the assumptions of constancy of the error variances and the independence are satisfied. If, on the other hand, the residuals of the new prediction equation show a clear pattern, then the experimenter should investigate a better prediction equation using quadratic terms. However, the two-level factorial designs are unable to provide us with the quadratic effects, and therefore we need to use other types of factorial designs, and this will be the objective of the next chapter.

Step 2: Once the constancy and the independence of the error terms are satisfied, the experimenter should construct the normal probability plot of the residuals. Attention should be given to the tails of the residuals. If the points that are on the tails fall close to the line, then normality of the error terms is satisfied.

If, on the other hand, the points in either tail exhibit a pattern similar to the ones that are presented in Figure 8.8 or 8.9, then the experimenter should define a new response Y^* where $Y^* = \sqrt{Y}$, or $Y^* = \log Y$, or $Y^* = \frac{1}{Y}$ and determine a new prediction equation. If the points fall approximately close to

the tails, then normality of the residuals is satisfied. In general, the transformation that does well in step 1 will also improve the normality of the error terms; i.e., only one transformation should be used to improve constancy, independence, and normality of the error terms.

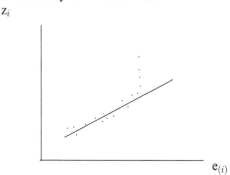

Figure 8.9. Departure from Normality (Right Tail)

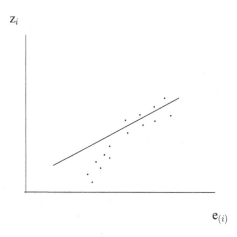

Figure 8.10. Departure from Normality (Left Tail)

Step 3: Once the constancy, the independence, and the normality of the error terms are satisfied, the experimenter should evaluate R_a^2, the adjusted coefficient of determination. If R_a^2 is at least .8, then at least 80% of the response is explained by the prediction equation.

If, on the other hand, the adjusted coefficient of determination is small, then the experimenter should investigate a better model using quadratic terms, and this will be the objective of Chapters 9 and 10.

Step 4: Finally, if replications are present, the experimenter should test the lack of fit. If there is no presence of lack of fit, then the prediction equation is valid.

If, on the other hand, there is strong evidence of lack of fit, then the experimenter should investigate a better model using quadratic terms, which will be presented in the next two chapters. Excellent references on regression analysis include Neter, Wasserman, and Kutner [1], Draper and Smith [2], Graybill [3], and Myers [4].

8.6 PROBLEMS

1. Consider the experimental design plan of problem 1 from Chapter 6.
   a. Does the prediction equation that relates the tensile strength with the important factors fit the data adequately?
   b. Does the prediction equation that relates the variability of the tensile strength with the important factors fit the data adequately?
   c. Is it possible to determine the factor settings that maximize the average strength and minimize the variability of the strength?
2. Consider the experimental design plan of problem 2 from Chapter 6.
   a. Does the prediction that relates the alertness with the important factors fit the data adequately?
   b. Determine the best prediction equation that relates the variability of the alertness with the important factors.
   c. Is it possible to determine the factor settings that maximize the alertness and minimize its variability?

REFERENCES

[1] Neter, J., Wasserman, W. and Kutner, M.H.   (1985), *Applied Linear Statistical Models*, 2nd edition, Richard D. Irwin, Inc., Homewood, IL.

[2] Draper, N.R. and Smith, H. (1966), *Applied Regression Analysis*, Wiley, New York.

[3] Graybill, F. (1976), *Introduction to the Theory of Linear Statistical Models*, Duxbury Press, Boston.

[4] Myers, R.H. (1990), *Classical and Modern Regression with Applications*, 2nd edition, Duxbury Press, Boston.

CHAPTER 9

Three-Level Factorial Designs

9.1 INTRODUCTION

There are situations for which designs other than the two-level factorial designs are necessary. Two such situations will be discussed in this chapter.

The first situation occurs when some factors are nonnumeric in which the level settings do not exist on a meaningful, continuous number scale. For instance, consider an experiment where one of the factors is the operator, and suppose that there are three operators. In this case, it is clear that designs other than the two-level factorial designs must be considered.

The second situation occurs when a linear relationship between the response and the factors is not feasible. As it was mentioned in Chapter 8, we recommend the investigation of a quadratic relationship. However, quadratic effects can be determined only when three-level designs are used.

Three-level designs are designs in which factors can take on three values, for example, low, medium, and high. Henceforth, for the sake of convenience and without loss of generality, we set the low values at -1, the medium values at 0, and the high values at $+1$. If the factors are numeric, then it is highly recommended that the factor settings be equally spaced within their respective ranges.

Next, we will present some of the most useful three-level factorial designs.

9.2. THREE-LEVEL FULL FACTORIAL DESIGN

Three-level full factorial designs are designs that test all three level combinations. A three-level full factorial design with n factors requires 3^n experimental runs to cover all possible combinations of the input factors.

This is illustrated in the factor columns in Table 9.1 for the case of two factors.

Table 9.1. The 3^2 experimental design plan

Run	A	B
1	+	+
2	0	+
3	−	+
4	+	0
5	0	0
6	−	0
7	+	−
8	0	−
9	−	−

Figure 9.1 below is a geometric representation of the three-level full factorial design with $n = 2$. As the figure shows, the design contains the two-level full factorial design points. These points are the corners of the square. It also contains the midpoints; these are the four points that are indicated by hollow rectangles. Finally, it also contains a center point, corresponding to the experimental run number 5 in which all factors are set at their medium level. This point is indicated by a shaded ellipse.

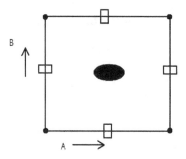

Figure 9.1. Graphical Representation of a 3^2 Full Factorial Design

Three-level full factorial designs are very easy to implement. We will illustrate this with an example.

Example 9.1: An experimenter needs to study the linear effects, the inter-action effects, and the quadratic effects of three numeric factors.

Since there are three factors, and the quadratic effects may be significant, each factor will be set at three levels. Consequently, a three-level full factorial design consists of $3^3 = 27$ experimental runs. The resulting design is illustrated in Table 9.2.

Table 9.2. Full factorial for three factors, each at three levels

	Factors				Factors				Factors		
Run	A	B	C	Run	A	B	C	Run	A	B	C
1	+	+	+	10	+	+	0	19	+	+	−
2	0	+	+	11	0	+	0	20	0	+	−
3	−	+	+	12	−	+	0	21	−	+	−
4	+	0	+	13	+	0	0	22	+	0	−
5	0	0	+	14	0	0	0	23	0	0	−
6	−	0	+	15	−	0	0	24	−	0	−
7	+	−	+	16	+	−	0	25	+	−	−
8	0	−	+	17	0	−	0	26	0	−	−
9	−	−	+	18	−	−	0	27	−	−	−

The three-level full factorial design in Table 9.2 is graphically displayed in Figure 9.2.

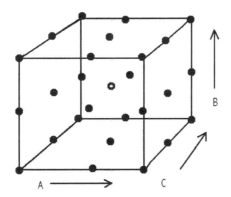

Figure 9.2. Graphical Representation of a 3^3 Full Factorial Design

The primary disadvantage of the three-level full factorial design is that the number of experimental runs is so large that experimenters can estimate unwanted high order interactions. For instance, the three-level full factorial design illustrated in Table 9.1 can be used to determine the main effects A and B; the linear two-factor interaction AB, the quadratic effects A^2 and B^2, and also the quadratic interaction terms such as A^2B and B^2A. The quadratic interaction terms will not be considered in the prediction equation; therefore, we need to implement designs cheaper than the three-level full factorial designs.

Notice that with only seven factors, a three-level full factorial design requires 3^7 or 2187 experimental runs. To reduce the total number of experimental runs, the use of fractional factorials in designs of three levels can be considered. Such designs are a bit more complicated, hence they shall not be covered in this book. A general procedure for constructing three-level fractional factorial designs is given in Montgomery [1], Connor and Zelen [2], and McLean and Anderson [3].

Next, we shall present a useful three-level factorial design called the Box-Behnken design.

9.3 BOX-BEHNKEN DESIGNS

Unlike the three-level full factorial designs, Box-Behnken designs are recommended only when the following two conditions are satisfied:
1. All three levels are evenly spaced.
2. There is no interest in corner points.

Box and Behnken [4] provided tabled Box-Behnken designs for up to sixteen factors, excluding the case of eight factors.

Letting k denote the number of quantitative factors, n_c denote the number of center runs, and N denote the total number of experimental runs, the construction of Box-Behnken designs varies depending on one of the following three situations:
1. $3 \leq k \leq 5$
2. $6 \leq k \neq 8 \leq 9$
3. $10 \leq k \leq 16$

We shall present in this section a few simple cases of Box-Behnken designs, and the remaining cases will be shown in Appendix 7.

Let us consider the case when there are three quantitative factors. The corresponding Box-Behnken design is illustrated in Table 9.3.

Table 9.3. A three-factor Box-Behnken design

Run	A	B	C
1	+	+	0
2	−	+	0
3	+	−	0
4	−	−	0
5	+	0	+
6	−	0	+
7	+	0	−
8	−	0	−
9	0	+	+
10	0	−	+
11	0	+	−
12	0	−	−
13	0	0	0
14	0	0	0
15	0	0	0

Note that the design shown in Table 9.3 uses a two-level full factorial design for two factors for each combination AB, BC, and AC while holding the remaining factor at the center, and also incudes experimental runs for all factors at the center. Since there are three combinations AB, BC, AC, there are three center point (0,0,0) runs.

The three-factor Box-Behnken design is graphically displayed in Figure 9.3.

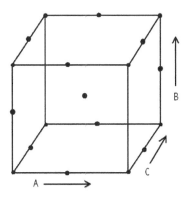

Figure 9.3. Graphical Representation of a Three-Factor Box-Behnken Design

Next, we will present the case when there are six quantitative factors. The shorthand notation is shown in Table 9.4.

Table 9.4. A six-factor Box-Behnken design

Run	A	B	C	D	E	F
1-8	± 1	± 1	0	± 1	0	0
9-16	0	± 1	± 1	0	± 1	0
17-24	0	0	± 1	± 1	0	± 1
25-32	± 1	0	0	± 1	± 1	0
33-40	0	± 1	0	0	± 1	± 1
41-48	± 1	0	± 1	0	0	± 1
49-54	0	0	0	0	0	0

The six-factor Box-Behnken design uses six blocks of two-level full factorial designs for three factors while maintaining the other three factors at the center. For instance, the first block of Table 9.4 consists of the following eight experimental runs.

Run	A	B	C	D	E	F
1	+	+	0	+	0	0
2	−	+	0	+	0	0
3	+	−	0	+	0	0
4	−	−	0	+	0	0
5	+	+	0	−	0	0
6	−	+	0	−	0	0
7	+	0	0	−	0	0
8	−	−	0	−	0	0

The six-factor Box-Behnken design also uses center runs. In fact, the number of required experimental runs at the center is $n_r = 6$.

Finally, we will also present a ten-factor Box-Behnken design. Its shorthand notation is illustrated in Table 9.5.

Table 9.5. A 10-factor Box-Behnken design

Run	A	B	C	D	E	F	G	H	J	K
1-16	0	± 1	0	0	0	± 1	± 1	0	0	± 1
17-32	± 1	± 1	0	0	± 1	0	0	0	0	± 1
33-48	0	± 1	± 1	0	0	0	± 1	± 1	0	0
49-64	0	± 1	0	± 1	0	± 1	0	0	± 1	0
65-80	± 1	0	0	0	0	0	0	± 1	± 1	± 1
81-96	0	0	± 1	± 1	± 1	0	0	0	0	± 1
97-112	± 1	0	0	± 1	0	0	± 1	± 1	0	0
113-128	0	0	± 1	0	± 1	0	± 1	0	± 1	0
129-144	± 1	0	± 1	0	0	± 1	0	0	± 1	0
145-160	0	0	0	± 1	± 1	± 1	0	± 1	0	0
161-170	0	0	0	0	0	0	0	0	0	0

The primary advantage of Box-Behnken designs is that the number of runs is much smaller than in the three-level full factorial design. Box-Behnken designs allow estimation of the main effects, the two-factor interaction effects, and all quadratic effects.

The primary disadvantage is the inability to predict the response at the extremes, that is, at the corners.

Next, we will present the most popular second-order design used by practitioners. It is called the central composite design.

9.4 CENTRAL COMPOSITE DESIGNS

The central composite designs (CCDs), attributed to Box and Wilson [5], are very efficient in determining main effects, two-factor interaction effects (if needed), and the quadratic effects.

The primary advantage of the central composite designs are their efficiency in minimizing the total number of experimental runs.

Because of their use in sequential experimentation, the central composite designs are the most popular second-order design. They involve the use of three separate parts that can be performed in a sequential manner.

The first part is the two-level factorial portion, which is often chosen to be a resolution V design. This factorial portion of resolution V contributes in estimating linear terms and two-factor interactions.

The second part is the center point portion. This portion contains the number of replications at the center. The number of replications at the center, denoted by n_c, is often calculated as

$$n_c = 4\sqrt{n_F + 1} - 2k,$$

where n_F is the number of runs in the factorial portion of the design, and k is the number of factors in the factorial portion of the design. The center points provide information about the existence of curvature in the system. If curvature is found in the system, then estimation of the quadratic terms will be determined through the use of the third part.

The third part is the axial portion. The axial portion contains the axial points, sometimes referred to as the star points. When there are k factors in the factorial portion, the axial portion contains $2k$ experimental runs. These are

x_1	x_2	$\cdots$	x_k
α	0	$\cdots$	0
$-\alpha$	0	$\cdots$	0
0	α	$\cdots$	0
0	$-\alpha$	$\cdots$	0
$\vdots$	$\vdots$	$\cdots$	$\vdots$
		$\cdots$	α
0	0	$\cdots$	$-\alpha.$

The axial points contribute to estimation of the quadratic terms. The choice of α depends to a great extent on the region of operability and region of interest. If the region of interest is larger than the region of operability, it is best to select the rotatable central composite design. We discuss this design next.

A. Rotatable Central Composite Design

The rotatable central composite design is a central composite design, where the axial distance $\alpha = (n_F)^{1/4}$ and n_F is the number of experimental runs in the factorial portion of the design. When a rotatable central composite design is used, α will take on values greater than 1, which means ± 1 no longer represents the factor minimum and maximum.

We illustrate the rotatable central composite design through the following example.

Example 9.2: The objective of a chemical engineer was to find settings of reaction time (A) and temperature (B) that produced maximum yield.

The region of operability that was determined by the chemical engineer is
A: Reaction time $= 30 - 40$ min
B: Temperature $= 150° - 160°F$
The rotatable central composite design is composed of three parts.

a. The first part is the two-level factorial portion; in this case, it is

Run	A	B
1	+	+
2	−	+
3	+	−
4	−	−

Hence $n_F = 4$.

b. The second part is the center point portion. It consists of n_c replications at the middle values of reaction time and temperature. $n_c = 4\sqrt{n_F + 1} - 2k$, $k = 2$, so

$$n_c = 4\sqrt{5} - 4 \simeq 5.$$

Hence, the second portion is

Run	A	B
5	0	0
6	0	0
7	0	0
8	0	0
9	0	0

c. The third part is the axial portion. Since $k = 2$, there are $2k = 4$ axial points. Here $\alpha = (n_F)^{1/4} = 1.414$.

Hence, the third portion is

Run	A	B
10	1.414	0
11	− 1.414	0
12	0	1.414
13	0	− 1.414

Putting all three portions together, the rotatable central composite design for two factors is illustrated in Table 9.6.

Table 9.6. Rotatable central composite design for two factors

Run	A	B
1	+	+
2	−	+
3	+	−
4	−	−
5	0	0
6	0	0
7	0	0
8	0	0
9	0	0
10	1.414	0
11	− 1.414	0
12	0	1.414
13	0	− 1.414

Before showing the graphical representation of Table 9.6, let us resolve the transformation to or from the design notation $\pm\,\alpha$,

$$\alpha = \frac{x - (\text{low} + \text{high})/2}{(\text{high} - \text{low})/2},$$

where x denotes the actual value and α is its corresponding coded value. The actual value is

$$x = \alpha \frac{(\text{high} - \text{low})}{2} + \frac{(\text{high} + \text{low})}{2}.$$

For example, setting the coded reaction time to $\alpha = 1.414$ is the same as setting the actual reaction time to

$$x = (1.414)\,\frac{(40 - 30)}{2} + \frac{(40 + 30)}{2},$$

$$x = 42.07,$$

and similarly, setting the coded reaction time to $-\alpha = -1.414$ is the same as setting the actual reaction time to

$$x = (-1.414)\frac{(40 - 30)}{2} + \frac{(40 + 30)}{2},$$

$$x = 27.93.$$

For the temperature, setting the coded temperature to $\alpha = 1.414$ is the same as setting the actual temperature to

$$x = 1.414 \times \frac{(160-150)}{2} + \frac{(150+160)}{2},$$

$$x = 162.07,$$

and similarly, setting the coded temperature to $-\alpha = -1.414$ is the same as setting the actual temperature to

$$x = -1.414 \frac{(160-150)}{2} + \frac{(150+160)}{2},$$

$$x = 147.93.$$

The rotatable central composite design for this example when the actual values of the reaction time and temperature are used is illustrated in Table 9.7.

Table 9.7. Rotatable central composite design with actual values

Run	A	B
1	40	160
2	30	160
3	40	150
4	30	150
5	35	155
6	35	155
7	35	155
8	35	155
9	35	155
10	42.07	155
11	27.93	155
12	35	162.07
13	35	147.93

For completeness, the graphical representation of Table 9.6 will appear in Figure 9.4.

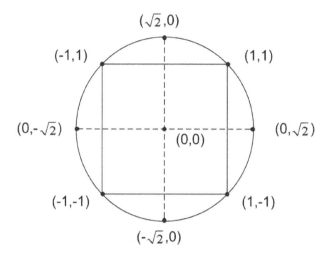

Figure 9.4. Central Composite Design for $k = 2$ and $\alpha = \sqrt{2}$.

Note that for $k = 2$, the design represents eight points equally spaced on a circle of radius $\sqrt{2}$, plus the center runs.

The central composite designs are more flexible than all other second-order designs. It allows the experimenter to choose any type of resolution plan for the factorial portion. This remark will be illustrated in the following two examples.

Example 9.3: An experimenter needs to study the relationship between three input factors and the output. He does not anticipate any two-factor interaction effects. He decides to use a rotatable central composite design.

The rotatable central composite design for this example ($k = 3$) is composed of the following three portions.

a. The first portion is the factorial portion. Since the experimenter does not anticipate any two-way interactions, a resolution III design is appropriate in saving runs.

Run	A	B	C
1	+	+	+
2	−	+	−
3	+	−	−
4	−	−	+

b. The second portion is the center point portion $n_c = 4\sqrt{n_F + 1} - 2k$. In our example, $k = 3$, and $n_F = 4$, hence

$$n_c = 4\sqrt{5} - 6 \approx 3.$$

The second portion is

Run	A	B	C
5	0	0	0
6	0	0	0
7	0	0	0

c. The third portion is the axial portion.
Since $k = 3$, there are $2k = 6$ axial points. Here $\alpha = (n_F)^{1/4} = 1.414$.
The third portion becomes

Run	A	B	C
8	1.414	0	0
9	− 1.414	0	0
10	0	1.414	0
11	0	− 1.414	0
12	0	0	1.414
13	0	0	− 1.414

Putting all three portions together, the resulting rotatable central composite design is illustrated in Table 9.8.

Table 9.8. Rotatable central composite design for three factors

Run	Factor		
	A	B	C
1	+	+	+
2	−	+	−
3	+	−	−
4	−	−	+
5	0	0	0
6	0	0	0
7	0	0	0
8	1.414	0	0
9	− 1.414	0	0
10	0	1.414	0

Table 9.8. (cont.)

	Factor		
Run	A	B	C
11	0	− 1.414	0
12	0	0	1.414
13	0	0	− 1.414

Another example that provides us with a rotatable central composite for three factors when the two-factor interactions are potentially important is given below.

Example 9.4: An experimenter needs to study the relationship between three input factors and the output. He anticipates the existence of two-factor interaction effects. He decides to construct a rotatable central composite design. It is composed of three portions.

a. The first portion is the factorial portion. Since the experimenter anticipates potential two-factor interactions, a resolution V design is chosen.

Run	A	B	C
1	+	+	+
2	−	+	+
3	+	−	+
4	−	−	+
5	+	+	−
6	−	+	−
7	+	−	−
8	−	−	−

b. The second portion is the center point portion

$$n_c = 4\sqrt{n_{F+1}} - 2k,$$

$$n_c = 4\sqrt{9} - 6 = 6.$$

The center point portion is

Run	A	B	C
9	0	0	0
10	0	0	0
11	0	0	0
12	0	0	0
13	0	0	0
14	0	0	0

c. The third portion is the axial portion with $k = 3$:

$$\alpha = (n_F)^{1/4} = 8^{1/4} = 2.828.$$

Run	A	B	C
15	2.828	0	0
16	− 2.828	0	0
17	0	2.828	0
18	0	− 2.828	0
19	0	0	2.828
20	0	0	− 2.828

Joining the three portions together, the resulting rotatable central composite design is illustrated in Table 9.9.

Table 9.9. Orthogonal central composite design for three factors

Run	A	B	C
1	+	+	+
2	−	+	+
3	+	−	+
4	−	−	+
5	+	+	−
6	−	+	−
7	+	−	−
8	−	−	−
9	0	0	0
10	0	0	0

Table 9.9. (cont.)

Run	A	B	C
11	0	0	0
12	0	0	0
13	0	0	0
14	0	0	0
15	2.828	0	0
16	− 2.828	0	0
17	0	2.828	0
18	0	− 2.828	0
19	0	0	2.828
20	0	0	− 2.828

The following table provides the most useful orthogonal central composite designs. Let n_A denote the number of axial points, and let N denote the total number of experimental runs.

Table 9.10. Values of α, n_F, n_A, n_f, and N for rotatable central composite design

# Factors	n_F	Resolution	n_A	n_c	α	N
2	4	III	4	5	1.414	13
3	4	III	6	3	1.414	13
3	8	Full	6	6	1.682	20
4	8	IV	8	4	1.682	20
4	16	Full	8	9	2	33
5	8	III	10	2	1.682	20
5	16	V	10	7	2	33
6	8	III	12	1	1.682	21
6	16	IV	12	5	2	33
6	32	VI	12	11	2.378	55
7	16	IV	14	3	2	33
7	64	VII	14	19	2.828	99
8	16	IV	16	1	2	33
8	64	V	16	17	2.828	97
9	32	IV	18	5	2.378	55
9	128	VI	18	28	3.363	174
10	32	IV	20	3	2.378	55
10	128	V	20	26	3.363	168

There are many practical situations in which scientists specify ranges for the factors, and these ranges are strict. That is, the region of interest and the region of operability are the same. When these situations occur, the face centered central composite design should be used.

B. Face Centered Central Composite Design

The face centered central composite is similar to the rotatable central composite design except for the fact that α is assigned a value of 1 instead of $(n_F)^{1/4}$.

For instance, suppose that there are two factors ($k = 2$) and that the region of interest and the region of operability are the same; then the face centered central composite design can easily be described as shown in Table 9.11.

Table 9.11. Face centered central composite design for $k = 2$

Run	A	B
1	+	+
2	−	+
3	+	−
4	−	−
5	0	0
6	0	0
7	0	0
8	0	0
9	0	0
10	1	0
11	− 1	0
12	0	1
13	0	− 1

The graphical representation of Table 9.11 is given in Figure 9.5.

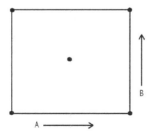

Figure 9.5. Face Centered Square (Central Composite Design with $\alpha = 1.0$) for $k = 2$.

Next, we shall present some useful three-level fractional factorial designs. They are called three-level Taguchi designs.

9.5 THREE-LEVEL TAGUCHI DESIGNS

The three-level Taguchi [6] designs are essentially a fractional portion of the three-level full factorial designs. These designs are mainly used when few of the two-factor interactions need to be investigated and few potential quadratic terms may be important. The three-level Taguchi designs are often used for qualitative factors.

The first three-level Taguchi design is presented in Table 9.12. The design is denoted as a $L_9(3^4)$. This is a design that requires nine experimental runs and can use up to four factors, where each factor has exactly three levels.

Table 9.12. L_9 design for three-level factors

Run #	1	2	3	4
1	-1	-1	-1	-1
2	-1	0	0	0
3	-1	1	1	1
4	0	-1	0	1
5	0	0	1	-1
6	0	1	-1	0
7	1	-1	1	0
8	1	0	-1	1
9	1	1	0	-1
Original factors and interations used to generate the matrix	a	b	$-ab$	ab^2

The next design that we shall present is the $L_{27}(3^{13})$. This is a design that requires 27 experimental runs and can use up to 13 factors, where each factor has exactly three levels. The $L_{27}(3^{13})$ is illustrated in Table 9.13.

Table 9.13. L_{27} design for three-level factors

No.	1	2	3	4	5	6	7	8	9	10
1	−	− 1	− 1	− 1	− 1	− 1	− 1	− 1	− 1	− 1
2	− 1	− 1	− 1	− 1	0	0	0	0	0	0
3	− 1	− 1	− 1	− 1	1	1	1	1	1	1
4	− 1	0	0	0	− 1	− 1	− 1	0	0	0
5	− 1	0	0	0	0	0	0	1	1	1
6	− 1	0	0	0	1	1	1	− 1	− 1	− 1
7	− 1	1	1	1	− 1	− 1	− 1	− 1	− 1	− 1
8	− 1	1	1	1	0	0	0	− 1	− 1	− 1
9	− 1	1	1	1	1	1	1	0	0	0
10	0	− 1	0	1	− 1	0	1	− 1	0	1
11	0	− 1	0	1	0	1	− 1	0	1	− 1
12	0	− 1	0	1	1	− 1	0	1	− 1	0
13	0	0	1	− 1	− 1	0	1	0	1	− 1
14	0	0	1	− 1	0	1	− 1	1	− 1	0
15	0	0	1	− 1	1	− 1	0	− 1	0	1
16	0	1	− 1	0	− 1	0	1	1	− 1	0
17	0	1	− 1	0	0	1	− 1	− 1	0	1
18	0	1	1	0	1	− 1	0	0	1	− 1
19	1	− 1	1	0	− 1	1	0	− 1	1	0
20	1	− 1	1	0	0	− 1	1	0	− 1	1
21	1	− 1	1	0	1	0	− 1	1	0	− 1
22	1	0	− 1	1	− 1	1	0	0	− 1	1
23	1	0	− 1	1	0	− 1	1	1	0	− 1
24	1	0	− 1	1	1	0	− 1	− 1	1	0
25	1	1	0	− 1	− 1	1	0	1	0	− 1
26	1	1	0	− 1	0	− 1	1	− 1	1	0
27	1	1	0	− 1	1	0	− 1	0	− 1	1
	a	b	− ab	ab^2	c	− ac	ac^2	− bc	abc	ab^{2}c^2

Table 9.13. (cont.)

No.	11	12	13
1	-1	-1	-1
2	0	0	0
3	1	1	1
4	1	1	1
5	-1	-1	-1
6	0	0	0
7	0	0	0
8	1	1	1
9	-1	-1	-1
10	-1	0	1
11	0	1	-1
12	1	-1	0
13	1	-1	0
14	-1	0	1
15	0	1	-1
16	0	1	-1
17	1	-1	0
18	-1	0	1
19	-1	1	0
20	0	-1	1
21	1	0	-1
22	1	0	-1
23	-1	1	0
24	0	-1	1
25	0	-1	1
26	1	0	-1
27	-1	1	0
	bc^2	$-ab^2c$	$-abc^2$

9.6 PROBLEMS

1. Discuss when you would use the following designs:
   * Three-level full factorial
   * Taguchi
   * Box-Behnken
   * Central composite
2. What three-level designs can you use for qualitative factors? For quantitative factors?

3. Construct a rotatable central composite design with five factors.

REFERENCES

[1] Montgomery, D.C. (1991), *Design and Analysis of Experiments*, 3rd edition, Wiley, New York.

[2] Connor, W.S. and Zelen, M. (1959), *Fractional Factorial Experimental Designs for Factors at Three Levels*, Applied Mathematics Series **54**, National Bureau of Standards, Washington, D.C.

[3] McLean, R.A. and Anderson, V.L. (1984), *Applied Factorial and Fractional Designs*, Marcel Dekker, New York.

[4] Box, G.E.P. and Behnken, D.W. (1960), Some new three level designs for the study of quantitative variables, *Technometrics* **2**, 455-475.

[5] Box, G.E.P. and Wilson, K.B. (1951), On the experimental attainment of optimum conditions, *J. Roy. Statist. Soc.* **B**:13, 1-38.

[6] Taguchi, G. and Konishi, S. (1987), *Taguchi Methods: Orthogonal Arrays and Linear Graphs*, American Supplier Institute, Inc., Dearborn, MI.

CHAPTER 10

Second-Order Analysis

10.1 INTRODUCTION

In many industrial experiments, it is necessary to analyze second-order effects on the response variable. This chapter deals with techniques that are effective in handling second-order models.

10.2 SECOND-ORDER MODEL IN MATRIX TERMS

A second-order model includes main effects, interactions, and quadratic effects.

For the case of p factors, the second-order model is

$$Y = \beta_0 + \beta_1 x_1 + \beta_2 x_2 \ldots + \beta_p x_p + \beta_{12} x_1 x_2 + \beta_{13} x_1 x_3 + \ldots$$

$$+ \beta_{p-1,p} x_{p-1} x_p + \ldots + \beta_{11} x_1^2 + \beta_{22} x_2^2 + \ldots + \beta_{pp} x_p^2 + \epsilon,$$

where ϵ is a random error term that satisfies the four conditions that were discussed in Chapter 8.

Second-order models are indeed a special case of the general linear regression model

$$Y = X\beta + \epsilon,$$

where

Y is a vector of observations
β is a vector parameter
X is a design matrix
ϵ is a vector of random errors.

This can be seen by letting

$$Y = \begin{bmatrix} Y_1 \\ Y_2 \\ \vdots \\ Y_n \end{bmatrix} \quad X = \begin{bmatrix} 1 & x_{11} \ldots x_{1p} & x_{11}x_{12} \ldots x_{1,p-1}x_{1,p} \ldots x_{11}^2 \ldots x_{1p}^2 \\ 1 & x_{21} \ldots x_{2p} & x_{21}x_{22} \ldots x_{2,p-1},x_{2,p} \ldots x_{21}^2 \ldots x_{2p}^2 \\ \vdots & \vdots & \vdots \\ 1 & x_{n1} \ldots x_{np} & x_{n1}x_{n2} \ldots x_{n,p-1},x_{n,p} \ldots x_{n1}^2 \ldots x_{np}^2 \end{bmatrix}$$

$$\beta = \begin{bmatrix} \beta_0 \\ \beta_1 \\ \vdots \\ \beta_p \\ \beta_{12} \\ \vdots \\ \beta_{p-1,p} \\ \beta_{11} \\ \vdots \\ \beta_{pp} \end{bmatrix} \qquad \epsilon = \begin{bmatrix} \epsilon_1 \\ \epsilon_2 \\ \vdots \\ \epsilon_n \end{bmatrix}.$$

For example, consider the second-order model with two factors:

$$Y = \beta_0 + \beta_1 x_1 + \beta_2 x_2 + \beta_{12} x_1 x_2 + \beta_{11} x_1^2 + \beta_{22} x_2^2 + \epsilon.$$

Let n denote the number of experimental runs, and let x_{ij} denote the ith experimental run taken at the jth factor. That is, $x_{i1}, \ldots, x_{ip}$ are the p settings that are taken when the ith experimental run is performed.

Let

$$Y = \begin{bmatrix} Y_1 \\ Y_2 \\ \vdots \\ Y_n \end{bmatrix},$$

where $Y_1, \ldots, Y_n$ are the n response values corresponding to the n experimental runs. Let

$$X = \begin{bmatrix} 1 & x_{11} & x_{12} & x_{11}x_{12} & x_{11}^2 & x_{12}^2 \\ 1 & x_{21} & x_{22} & x_{21}x_{22} & x_{21}^2 & x_{22}^2 \\ \vdots & \vdots & \vdots & \vdots & \vdots & \vdots \\ 1 & x_{n1} & x_{n2} & x_{n1}x_{n2} & x_{n1}^2 & x_{n2}^2 \end{bmatrix}$$

$$\beta = \begin{bmatrix} \beta_0 \\ \beta_1 \\ \beta_2 \\ \beta_{12} \\ \beta_{11} \\ \beta_{22} \end{bmatrix} \qquad \epsilon = \begin{bmatrix} \epsilon_1 \\ \epsilon_2 \\ \vdots \\ \epsilon_n \end{bmatrix}.$$

Using Appendix 8, it follows that the second-order model with two factors can be expressed as

$$Y = X\beta + \epsilon.$$

A simpler representation of the second-order model with two factors is obtained by letting

$$x_3 = x_1 x_2, \ x_4 = x_1^2, \ x_5 = x_2^2,$$

$$\beta_3 = \beta_{12}, \beta_4 = \beta_{11}, \text{ and } \beta_5 = \beta_{22}.$$

The second-order model becomes

$$Y = \beta_0 + \beta_1 x_1 + \beta_2 x_2 + \beta_3 x_3 + \beta_4 x_4 + \beta_5 x_5 + \epsilon,$$

which is clearly a linear regression model.

10.3 ESTIMATION OF THE SECOND-ORDER MODEL PARAMETERS

Let us denote the vector of estimated regression coefficients

$$b_0, b_1, \ldots, b_{pp} \text{ as } b:$$

$$b = \begin{bmatrix} b_0 \\ b_1 \\ \vdots \\ b_p \\ b_{12} \\ \vdots \\ b_{p-1,p} \\ b_{11} \\ \vdots \\ b_{pp} \end{bmatrix},$$

and let the estimated regression equation be

$$\widehat{Y} = b_0 + b_1 x_1 + b_2 x_2 + \ldots + b_p x_p$$

$$+ b_{12} x_1 x_2 + b_{13} x_1 x_3 + \ldots + b_{p-1,p} x_{p-1} x_p$$

$$+ b_{11} x_1^2 + b_{22} x_2^2 + \ldots + b_{pp} x_p^2,$$

which can also be represented as

$$\widehat{Y} = Xb.$$

Note that the second-order model has $2p + \frac{p(p-1)}{2} + 1$ terms.
The vector b is determined with the objective of minimizing $e'e$, where

$$e = Y - \widehat{Y}.$$

It was determined that the best estimated regression coefficients are

$$b = (X'X)^{-1}X'Y.$$

10.4 ESTIMATION OF THE FIRST-ORDER MODEL PARAMETERS

Next, we shall apply the above formula to determine the estimated regression equation of a first-order model without interactions.
Let

$$\widehat{Y} = b_0 + b_1 x_1 + b_2 x_2 + \ldots + b_p x_p.$$

Let n denote the number of experimental runs of a balanced and orthogonal factorial design. Such designs include two-level fractional factorial designs and two-level full factorial designs.
The design matrix is

$$X = \begin{bmatrix} 1 & x_{11} & \ldots & x_{1p} \\ 1 & x_{12} & \ldots & x_{2p} \\ \vdots & & & \\ 1 & x_{n1} & \ldots & x_{np} \end{bmatrix},$$

where $x_{ij} = \pm 1$ for all $i = 1, \ldots, n$ and $j = 1, \ldots, p$, $\sum_{i=1}^{n} x_{ij} = 0$ for all $j = 1, \ldots, p$ (balanced design), and

$$\sum_{i=1}^{n} x_{ij} x_{ik} = 0 \text{ for all } j \neq k \text{ (orthogonality)}.$$

Using Appendix 8, the estimated regression coefficients are the products of the matrix $(X'X)^{-1}$ and the vector $(X'Y)$.
The inverse of the matrix $X'X$, denoted by $(X'X)^{-1}$, is

$$(\mathbf{X'X})^{-1} = \left(\begin{bmatrix} 1 & 1 & \cdots & 1 \\ x_{11} & x_{21} & \cdots & x_{n1} \\ x_{12} & & & \vdots \\ \vdots & \vdots & & \vdots \\ x_{1p} & x_{2p} & & x_{np} \end{bmatrix} \begin{bmatrix} 1 & x_{11} & \cdots & x_{1p} \\ 1 & x_{21} & \cdots & x_{2p} \\ \vdots & \vdots & & \\ \vdots & \vdots & & \\ 1 & x_{n1} & \cdots & x_{np} \end{bmatrix} \right)^{-1}$$

$$= \begin{bmatrix} n & 0 & 0 & 0 & 0 \\ 0 & n & 0 & 0 & 0 \\ 0 & 0 & \ddots & 0 & 0 \\ 0 & 0 & 0 & \ddots & 0 \\ 0 & 0 & 0 & 0 & n \end{bmatrix}^{-1}$$

$$= \tfrac{1}{n} I,$$

where I is the identity matrix with n rows and n columns.
The vector $\mathbf{X'Y}$ is

$$\mathbf{X'Y} = \begin{bmatrix} 1 & 1 & \cdots & 1 \\ x_{11} & x_{21} & \cdots & x_{n1} \\ \vdots & & & \\ x_{1p} & x_{2p} & & x_{np} \end{bmatrix} \begin{bmatrix} Y_1 \\ Y_2 \\ \vdots \\ Y_n \end{bmatrix}$$

$$= \begin{bmatrix} \sum_{i=1}^{n} Y_i \\ \sum_{i=1}^{n} x_{i1} Y_i \\ \vdots \\ \sum_{i=1}^{n} x_{ip} Y_i \end{bmatrix}.$$

Therefore,

$$b = \begin{bmatrix} \dfrac{\sum\limits_{i=1}^{n} Y_i}{n} \\[2ex] \dfrac{\sum\limits_{i=1}^{n} x_{i1} Y_i}{n} \\[2ex] \dfrac{\sum\limits_{i=1}^{n} x_{ij} Y_i}{n} \\[2ex] \dfrac{\sum\limits_{i=1}^{n} x_{ip} Y_i}{n} \end{bmatrix} .$$

Now, using the fact that $\frac{n}{2}$ experimental runs are taken at the high level $(+1)$ and $\frac{n}{2}$ are taken at the low level (-1), then

$$\frac{\sum\limits_{i=1}^{n} x_{ij} Y_i}{n} = \frac{\sum\limits^{n/2} Y_i^+}{n} - \frac{\sum\limits^{n/2} Y_i^-}{n},$$

where Y_i^+ is the ith response value taken at the high level $(+1)$ of the factor x_i and Y_i^- is the ith response value taken at the low level (-1) of the factor x_i.
Note that

$$\frac{\sum\limits^{n/2} Y_i^+}{n} - \frac{\sum\limits^{n/2} Y_i^-}{n} = \frac{1}{2}\left(\frac{\sum\limits^{n/2} Y_i^+}{\frac{n}{2}} - \frac{\sum\limits^{n/2} Y_i^-}{\frac{n}{2}} \right),$$

which is exactly half the effect of the factor x_i. The estimated regression equation becomes

$$\widehat{Y} = \overline{Y} + \frac{\Delta_1}{2} x_1 + \ldots + \frac{\Delta_j}{2} x_j + \ldots \frac{\Delta_p}{2} x_p,$$

where Δ_j; $j = 1, \ldots, p$ is half the effect of factor x_j.
The above estimated regression equation agrees with the prediction equation that we routinely used in the earlier chapters.

10.5 FITTING A SECOND-ORDER MODEL

In this section, we will present an application of the second-order analysis to a three-level full factorial design.
Example 10.1: An experimenter wishes to determine a second-order model that relates carbon monoxide in g/m^3 with two factors. Factor A is the amount of ethanol added to a standard fuel and factor B represents the air/fuel ratio. The experimenter decides to use a three-level full factorial design plan as shown in Table 10.1.

Table 10.1. Three-level full factorial design

Runs	X_1	X_2	Y
1	1	1	59
2	0	1	68
3	-1	1	67
4	1	0	77
5	0	0	80
6	-1	0	70
7	1	-1	92
8	0	-1	80
9	-1	-1	64

The experimenter wants to determine the estimated regression equation

$$\widehat{Y} = b_0 + b_1 x_1 + b_2 x_2 + b_{12} x_1 x_2 + b_{11} x_1^2 + b_{22} x_2^2.$$

The design matrix X for this model is

$$X = \begin{bmatrix} & x_1 & x_2 & x_1 x_2 & x_1^2 & x_2^2 \\ 1 & 1 & 1 & 1 & 1 & 1 \\ 1 & 0 & 1 & 0 & 0 & 1 \\ 1 & -1 & 1 & -1 & 1 & 1 \\ 1 & 1 & 0 & 0 & 1 & 0 \\ 1 & 0 & 0 & 0 & 0 & 0 \\ 1 & -1 & 0 & 0 & 1 & 0 \\ 1 & 1 & -1 & -1 & 1 & 1 \\ 1 & 0 & -1 & 0 & 0 & 1 \\ 1 & -1 & -1 & 1 & 1 & 1 \end{bmatrix},$$

and the response vector is

$$Y = \begin{bmatrix} 59 \\ 68 \\ 67 \\ 77 \\ 80 \\ 70 \\ 92 \\ 80 \\ 64 \end{bmatrix}.$$

Note that the entries in the columns associated with x_1^2 and x_2^2 are found by squaring the entries in column x_1 and x_2, respectively, and the entries in the $x_1 x_2$ column are found by multiplying each entry from x_1 by the corresponding entry from x_2.

The $X'X$ matrix is

$$
\begin{bmatrix}
9 & 0 & 0 & 0 & 6 & 6 \\
0 & 6 & 0 & 0 & 0 & 0 \\
0 & 0 & 6 & 0 & 0 & 0 \\
0 & 0 & 0 & 4 & 0 & 0 \\
6 & 0 & 0 & 0 & 6 & 4 \\
6 & 0 & 0 & 0 & 4 & 6
\end{bmatrix},
$$

its inverse $(X'X)^{-1}$ is

$$
\begin{bmatrix}
\frac{5}{9} & 0 & 0 & 0 & -\frac{1}{3} & -\frac{1}{3} \\
0 & \frac{1}{6} & 0 & 0 & 0 & 0 \\
0 & 0 & \frac{1}{6} & 0 & 0 & 0 \\
0 & 0 & 0 & \frac{1}{4} & 0 & 0 \\
0 & 0 & 0 & 0 & \frac{1}{2} & 0 \\
0 & 0 & 0 & 0 & 0 & \frac{1}{2}
\end{bmatrix},
$$

$$
X'Y =
\begin{bmatrix}
657 \\
27 \\
-42 \\
-36 \\
429 \\
430
\end{bmatrix},
$$

and from $b = (X'X)^{-1}X'Y$ we obtain

$$
b =
\begin{bmatrix}
78.6 \\
4.5 \\
-7.0 \\
-9.0 \\
-4.5 \\
-4.0
\end{bmatrix}.
$$

Therefore the fitted quadratic model for carbon monoxide is

$$\widehat{Y} = 78.6 + 4.5x_1 - 7x_2 - 9x_1x_2 - 4.5x_1^2 - 4x_2^2.$$

Although it appears that every coefficient is different from 0, we should determine whether the coefficients are significantly different from zero.

10.6 INFERENCES ABOUT REGRESSION PARAMETERS

This section presents a formal approach that enables the user to decide which main effects are important, which two-factor interactions are important, and which quadratic terms are important.

The solution will be given for a general case. That is, for L terms and among these L terms, there could be main effects only, main effects with the two-factor interactions, main effects without interactions but with quadratic terms, or main effects with the two-factor interactions and with quadratic terms also.

Let b_k be the kth regression coefficient that was determined from the equation

$$b = (X'X)^{-1}X'Y,$$

where

$$X = \begin{pmatrix} 1 & x_{11} & \ldots & x_{1L} \\ 1 & x_{21} & \ldots & x_{2L} \\ \vdots & \vdots & \vdots & \vdots \\ 1 & x_{n1} & \ldots & x_{nL} \end{pmatrix}$$

and

$$Y = \begin{pmatrix} Y_1 \\ \vdots \\ Y_n \end{pmatrix}.$$

Recall that in the above design matrix, we allow possibilities such as

$$x_k = x_i x_j \quad \text{or} \quad x_k = x_i^2.$$

In order to determine whether the term x_k should go into the estimated prediction equation

$$\widehat{Y} = b_0 + \sum_{i=1}^{L} b_i x_i,$$

the experimenter needs to test the hypothesis $H_0 : \beta_k = 0$ versus $H_a : \beta_k \neq 0$,

and this can be achieved by the following three steps.

Step 1: Compute

$$T_0 = \frac{b_k}{S(b_k)},$$

where $S^2(b_k)$ is the kth diagonal element of the estimated variance–covariance matrix $S^2(b)$. Calculations show that the estimated variance–covariance matrix of the vector

$$b = \begin{pmatrix} b_0 \\ b_1 \\ \vdots \\ b_L \end{pmatrix}$$

is

$$S^2(b) = MSE(\mathrm{X}^t\mathrm{X})^{-1},$$

where MSE $= \dfrac{\sum\limits_{i=1}^{n} e_i^2}{n-L-1}$.

Step 2: Determine T* from Appendix 5, where

$$\mathrm{T}^* = t(\alpha/2, n - L - 1),$$

and

$$\alpha = 0.05 \text{ or } 0.01.$$

Step 3: Compare $|T_0|$ to T*. If $|T_0| > \mathrm{T}^*$ then we reject the null hypothesis. The term x_k should be included in the regression equation. If $|T_0| \leq \mathrm{T}^*$, then we accept H_0. The term x_k should be excluded from the regression model.

10.7 CONFIDENCE LIMITS FOR PREDICTED VALUES

Consider the general linear model

$$\mathrm{Y} = \mathrm{X}\beta + \epsilon,$$

where

$$X = \begin{pmatrix} 1 & x_{11} & \cdots & x_{1L} \\ 1 & x_{21} & \cdots & x_{2L} \\ \vdots & \vdots & \vdots & \vdots \\ 1 & x_{n1} & \cdots & x_{nL} \end{pmatrix},$$

$$Y = \begin{pmatrix} Y_1 \\ \vdots \\ Y_n \end{pmatrix},$$

and the regression parameters

$$\beta = \begin{pmatrix} \beta_0 \\ \beta_1 \\ \vdots \\ \beta_L \end{pmatrix}$$

are estimated by

$$b = (X'X)^{-1}X'Y.$$

Let X_0 be a new experimental run; then its predicted response $\widehat{Y}_0$ is

$$\widehat{Y}_0 = X_0'b.$$

Let the mean square error (MSE) be determined as in Section 10.5.
The $100(1 - \alpha)$ percent confidence interval for the predicted response $\widehat{Y}_0$ at the point X_0 is an interval (a, b), where

$$a = \widehat{Y}_0 - t(\alpha/2, n - L - 1)\sqrt{MSE}\sqrt{1 + X_0'(X'X)^{-1}X_0}$$

and

$$b = \widehat{Y}_0 + t(\alpha/2, n - L - 1)\sqrt{MSE}\sqrt{1 + X_0'(X'X)^{-1}X_0}.$$

Here, $t(\alpha/2, n - L - 1)$ denotes the critical value of the student distribution with $n - L - 1$ degrees of freedom, n denotes the number of observations, and L denotes the number of terms excluding β_0.

Example 10.2: We wish to determine the 95% confidence interval for a new experimental run taken at $(1/2, -1)$.

Suppose that $m = 2$ and that a model with main effects, interaction terms, and quadratic terms is a good model. That is, suppose that

$$\widehat{Y} = b_0 + b_1 x_1 + b_2 x_2 + b_{12} x_1 x_2 + b_{11} x_1^2 + b_{22} x_2^2.$$

Estimated coefficients are $b_0, b_1, b_2, b_{12}, b_{11},$ and b_{22} and are determined from Section 10.2.

The 95% prediction interval at the new experimental run $(1/2, -1)$ will be derived as follows:

1. Replace x_1 by $+1/2$ and x_2 by -1, and then evaluate $\widehat{Y}_0$.
2. Determine the MSE.
3. Construct the design matrix

$$
\begin{pmatrix}
1 & x_{11} & x_{12} & x_{11}x_{12} & x_{11}^2 & x_{12}^2 \\
1 & x_{21} & x_{22} & x_{21}x_{22} & x_{21}^2 & x_{22}^2 \\
\vdots & \vdots & \vdots & \vdots & \vdots & \vdots \\
1 & x_{n1} & x_{n2} & x_{n1} & x_{n1}^2 & x_{n2}^2
\end{pmatrix}.
$$

4. Determine X_0:

$$
X_0 = \begin{pmatrix}
1 \\
x_{01} \\
x_{02} \\
x_{03} \\
x_{04} \\
x_{05}
\end{pmatrix},
$$

where

$$
x_{01} = 1/2
$$

$$
x_{02} = -1
$$

$$
x_{03} = (1/2) \times -1 = -1/2
$$

$$
x_{04} = (1/2)^2 = 1/4
$$

$$
x_{05} = (-1)^2 = 1.
$$

5. Determine $X_0'(X'X)^{-1}X_0$.
6. Determine $t(.025, n-6)$.
7. Find the lower bound a and the upper bound b,

$$
a = \widehat{Y}_0 + (\alpha/2, n-6)\sqrt{MSE}\sqrt{1 + X_0'(X'X)^{-1}X_0}
$$

and

$$
b = \widehat{Y}_0 + (\alpha/2, n-6)\sqrt{MSE}\sqrt{1 + X_0'(X'X)^{-1}X_0}.
$$

10.8 VALIDITY OF THE PREDICTION EQUATION

In Chapter 8, we discussed graphical methods and formal statistical tests for studying the validity of the prediction equation. The same methods can be used to check the validity of the quadratic prediction equation.
For example, let $p = 3$; then the quadratic regression equation is

$$Y = \beta_0 + \beta_1 x_1 + \beta_2 x_2 + \beta_3 x_3 + \beta_{12} x_1 x_2 + \beta_{13} x_1 x_3 + \beta_{23} x_2 x_3$$

$$+ \beta_{11} x_1^2 + \beta_{22} x_2^2 + \beta_{33} x_3^2 + \epsilon.$$

A simple change of variable such as

$$x_1 x_2 = x_4, x_1 x_3 = x_5, x_2 x_3 = x_6$$

$$x_1^2 = x_7, x_2^2 = x_8, x_3^2 = x_9$$

and

$$\beta_{12} = \beta_4, \beta_{13} = \beta_5, \beta_{23} = \beta_6, \beta_{11} = \beta_7, \beta_{22} = \beta_8, \beta_{33} = \beta_9$$

reduces the quadratic model into

$$Y = \beta_0 + \beta_1 x_1 + \beta_2 x_2 + \beta_3 x_3 + \beta_4 x_4 + \beta_5 x_5 + \beta_6 x_6$$

$$+ \beta_7 x_7 + \beta_8 x_8 + \beta_9 x_9 + \epsilon,$$

and the latter model can be treated as if there were nine input factors and only their main effects might be significant.

10.9 QUADRATIC OPTIMIZATION

Let $b_0, b_1, \ldots, b_m, \ldots, b_{m-1,m}, b_{11}, \ldots, b_{mm}$ be the estimated regression parameters that were determined in Section 10.2. If the fitted regression equation fits the data adequately, then the model

$$\hat{Y} = b_0 + b_1 x_1 + \ldots b_m x_m + b_{12} x_1 x_2 + \ldots b_{1m} x_1 x_m + \ldots b_{m-1,m} x_{m-1} x_m$$

$$+ b_{11} x_1^2 + \ldots b_{mm} x_m^2$$

should be used to determine if the surface has a minimum, a maximum, or a saddle point.
The fitted regression equation can be rewritten as

$$\widehat{Y} = b_0 + X'b + X'BX,$$

where

$$X = \begin{pmatrix} x_1 \\ x_2 \\ \vdots \\ x_m \end{pmatrix},$$

$$b = \begin{pmatrix} b_1 \\ b_2 \\ \vdots \\ b_m \end{pmatrix},$$

and

$$B = \begin{pmatrix} b_{11} & b_{12}/2 & \cdots & b_{1m}/2 \\ b_{12}/2 & b_{22} & \cdots & b_{2m}/2 \\ \vdots & \vdots & \ddots & \vdots \\ b_{1m}/2 & b_{2m}/2 & \cdots & b_{mm} \end{pmatrix}.$$

Note 10.1: The matrix B is a symmetric m by m matrix.

A. Stationary Point

The stationary (critical) point is the point for which the first derivative of Y with respect to the vector X is equal to zero. That is,

$$\frac{dY}{dX} = 0.$$

However, matrix differentiation gives

$$\frac{dY}{dX} = b + 2BX.$$

Hence, the stationary point X_0 is

$$X_0 = -\tfrac{1}{2}B^{-1}b.$$

B. Optimal Point

The optimal point depends on the sign of the eigenvalues of the matrix B.
Recall that the eigenvalues of a matrix B are the roots $\lambda_1, \lambda_2, \ldots, \lambda_m$ of the equation

$$\text{Determinant of } (B - \lambda I) = 0,$$

where λ is a real number, and I is the identity matrix.
These are the following cases:
1. If all eigenvalues are positive, then X_0 is a minimum point.
2. If all eigenvalues are negative, then X_0 is a maximum point.
3. If some eigenvalues are positive and others are negative, then X_0 is a saddle point.

Note 10.2: After determining X_0, the optimal Y_0 can be determined by substituting X_0 in the equation

$$Y_0 = b_0 + X_0'b + X_0'BX_0.$$

Example 10.3: Suppose that $m = 2$; then

$$\widehat{Y} = b_0 + b_1x_1 + b_2x_2 + b_{12}x_1x_2 + b_{11}x_1^2 + b_{22}x_2^2.$$

In a matrix form

$$Y = b_0 + X^t b + X^t BX,$$

where

$$X = \begin{pmatrix} x_1 \\ x_2 \end{pmatrix}$$

$$b = \begin{pmatrix} b_1 \\ b_2 \end{pmatrix}$$

$$B = \begin{pmatrix} b_{11} & b_{12}/2 \\ b_{12}/2 & b_{22} \end{pmatrix}.$$

If Y is a stationary point at X_0, then

$$\frac{dY}{dX_1} = 0 \text{ and } \frac{dY}{dX_2} = 0.$$

That is,

$$b_1 + b_{12}x_2 + 2b_{11}x_1 = 0$$

$$b_2 + b_{12}x_1 + 2b_{22}x_2 = 0.$$

In a matrix form

$$b + 2BX = 0.$$

The above system of equations or its matrix representation is equivalent to

$$b_{11}x_1 + \frac{b_{12}}{2}x_2 = -b_1/2$$

$$b_{22}x_2 + \frac{b_{12}}{2}x_1 = -b_2/2.$$

In matrix form,

$$\begin{pmatrix} b_{11} & b_{12}/2 \\ b_{12}/2 & b_{22} \end{pmatrix} \begin{pmatrix} X_1 \\ X_2 \end{pmatrix} = \begin{pmatrix} -b_1/2 \\ -b_2/2 \end{pmatrix} = -b/2.$$

Let Δ denote the determinant of the matrix B; then

$$\Delta = b_{11}b_{22} - \frac{b_{12}^2}{4}.$$

If $\Delta \neq 0$, then the stationary point is given by

$$X_0 = B^{-1}b,$$

which is

$$X_0 = \frac{1}{\Delta} \begin{pmatrix} b_{22} & -b_{12}/2 \\ -b_{12}/2 & b_{11} \end{pmatrix} \begin{pmatrix} -b_1/2 \\ -b_2/2 \end{pmatrix}.$$

The optimal point will be determined by finding the eigenvalues of the matrix B, that is, by solving

$$\text{Determinant of } (B - \lambda I) = 0,$$

where

$$\lambda = \begin{pmatrix} \lambda_1 \\ \lambda_2 \end{pmatrix}$$

and

$$I = \begin{pmatrix} 1 & 0 \\ 0 & 1 \end{pmatrix}.$$

If λ_1 and λ_2 are positive, then X_0 is a minimum. If λ_1 and λ_2 are negative, then X_0 is a maximum. If λ_1 and λ_2 are of opposite signs, then X_0 is a saddle point.

More details of quadratic optimization are given in several experimental design books such as those by Box and Draper [1], Box, Hunter, and Hunter [2], Davis [3], Khuri and Cornell [4], Montgomery [5], and Myers and Montgomery [6].

We conclude this book with a general flow chart (Figure 10.1) that provides the experimenter with some possible paths that can be taken for process optimization.

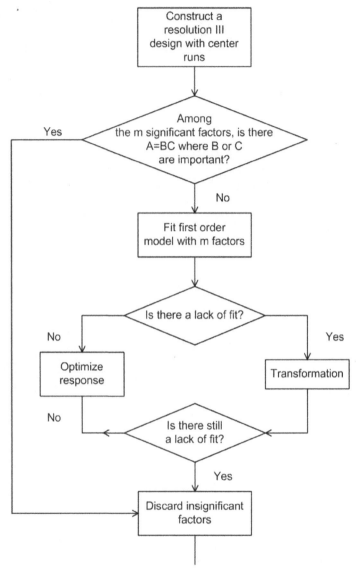

Figure 10.1. Sequential Approach for Design Optimization

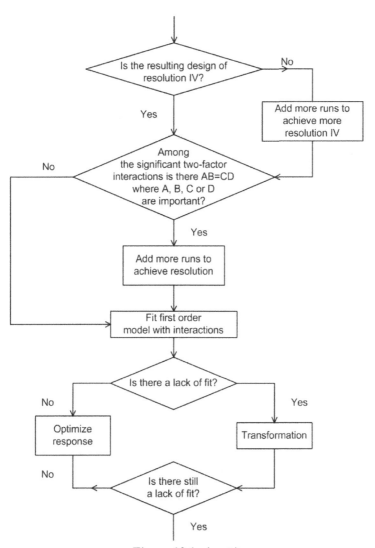

Figure 10.1. (cont.)

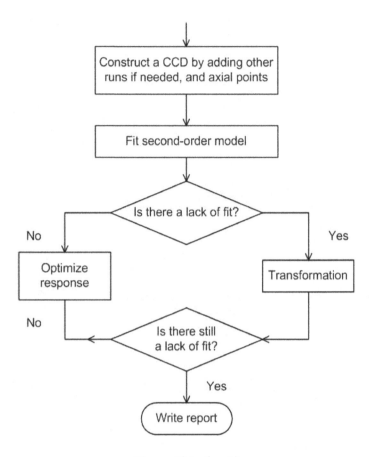

Figure 10.1. (cont.)

10.10 PROBLEMS

1. Use matrix algebra to determine the estimated regression equation for a general first-order model with interactions. Assume that the two-level factorial design is balanced and orthogonal.
2. A chemical engineer is interested in determining the operating conditions that maximize the yield of a process. Two factors are considered, reaction time and reaction temperature. The experimenter expects that the quadratic effects may be important. He decides to use a rotatable central composite design, as we did in Section 9.3.1. He obtains the following data:

Coded	Variables	Response
x_1	x_2	y
-1	-1	76.5
-1	1	77.0
1	-1	78.0
1	1	79.5
0	0	79.9
0	0	80.3
0	0	80.0
0	0	79.7
0	0	79.8
1.414	0	78.4
-1.414	0	75.6
0	1.414	78.5
0	-1.414	77.0

a. Determine the estimated quadratic regression equation.
b. Determine whether the quadratic effects are statistically significant or not. Use $\alpha = .05$.
c. Determine a 90% confidence interval for the predicted response $\hat{Y}_0$ at $x_1 = \frac{1}{2}$ and $x_2 = \frac{1}{2}$.

REFERENCES

[1] Box, G.E.P. and Draper, N.R. (1987), *Empirical Model Building and Response Surfaces*, Wiley, New York.

[2] Box, G.E.P., Hunter, W.G. and Hunter, J.S. (1978), *Statistics for Experimenters*, Wiley, New York.

[3] Davis, O.L.. (1956), *Design and Analysis of Industrial Experiments*, 2nd edition, Hafner Publishing Co., New York.

[4] Khuri, A.I. and Cornell, J.A. (1987), *Response Surfaces: Designs and Analyses of Variance*, Marcel Dekker, New York.

[5] Montgomery, D.C. (1991), *Design and Analysis of Experiments*, 3rd edition, Wiley, New York.

[6] Myers, R.H. and Montgomery, D.C. (1995), *Response Surface Methodology*, Wiley, New York.

Two-Level Fractional Factorial Designs

This appendix will provide us with the most useful two-level fractional factorial designs of resolution III, IV and V. It shall be used as a cookbook approach.

Number of factors	Resolution	# of runs	Generators
3	III	4	C = AB
4	IV	8	D = ABC
5	V	16	E = ABCD
	III	8	D = AB
			E = AC
6	VI	32	F = ABCDE
	IV	16	E = ABC
			F = BCD
	III	8	D = AB
			E = AC
			F = BC
7	VII	64	G = ABCDEF
	IV	32	F = ABCD
			G = ABDE
	IV	16	E = ABC
			F = BCD
			G = ACD
	III	8	D = AB
			E = AC
			F = BC
			G = ABC
8	V	64	G = ABCD
			H = ABEF
	IV	32	F = ABC
			G = ABD
			H = BCDE
	IV	16	E = BCD
			F = ACD
			G = ABC
			H = ABD
9	VI	128	H = ACDFG
			J = BCEFG
	IV	64	G = ABCD

			H = ACEF
			J = CDEF
	IV	32	F = BCDE
			G = ACDE
			H = ABDE
			J = ABCE
	III	16	E = ABC
			F = BCD
			G = ACD
			H = ABD
			J = ABCD
10	V	128	H = ABCG
			J = ACDE
			K = ACDF
	IV	64	G = BCDF
			H = ACDF
			J = ABDE
			K = ABCE
	IV	32	F = ABCD
			G = ABCE
			H = ABDE
			J = ACDE
			K = BCDE
	III	16	E = ABC
			F = BCD
			G = ACD
			H = ABD
			J = ABCD
			K = AB

Next, an example is given that shows how this table should be used.

Example: Consider an experiment with six factors to construct a design of resolution III. The table shows that eight experimental runs are needed. Hence we will construct a full factorial at two levels with three factors A, B, and C.

A	B	C
+	+	+
−	+	+
+	−	+
−	−	+
+	+	−
−	+	−
+	−	−
−	−	−

Then complete the experimental design plan by constructing the fourth column as the product of the first two columns,

$$D = AB,$$

the fifth column as the product of the first column and the third column,

$$E = AC,$$

and the sixth column as the product of the second column and the third column

$$F = BC.$$

The resulting experimental plan is

Run	A	B	C	D	E	F
1	+	+	+	+	+	+
2	−	+	+	−	−	+
3	+	−	+	−	+	−
4	−	−	+	+	−	−
5	+	+	−	+	−	−
6	−	+	−	−	+	−
7	+	−	−	−	−	+
8	−	−	−	+	+	+

Plackett-Burman Designs

This appendix will provide us with Plackett-Burman designs.

Number of factors	Number of runs	Generator
4-7	8	+++ − + − −
8-11	12	++ − +++ − − − + −
12-15	16	++++ − + − ++ − − + − − −
16-19	20	++ − − ++++ − + − + − − − − − ++ −
20-23	24	+++++ − + − ++ − − ++ − − + − + − − − −
24-31	36	− + − +++ − − − − +++++ − +++ − − + − − − − + − + − ++ − − + −
32-63	64	++++++ − + − + − ++ − − ++ − +++ − ++ − + − − + − − +++ − − − + − ++++ − − + − + − − − ++ − − − − + − − − − −
64-127	128	+++++++ − + − + − + − − ++ − − +++ − +++ − + − − + − ++ − − − ++ − ++++ − ++ − + − ++ − ++ − − + − − + − − − + ++ − − − − + − +++++ − − + − + − +++ − − ++ − + − − − − + − − ++ + − − − − + − + − − − − − ++ − − − − − + − − − − − −

Next, an example is given to show the construction of Plackett-Burman designs.

Example: Consider an experiment with six factors. To construct a Plackett-Burman design, the table shows that eight experimental runs are needed. The construction will be as follows. The single generating vector from the task is +++ − + − − .

1. Use the generating vectors as column A.
2. Build column B by making the last value of A the first value in B, and then slide the rest of the A values below that value.

3. Build column C by making the last value of B the first value in C, and then slide the rest of the B values below that value.
4. Continue until all columns are complete, and add − at the last row.

Run	A	B	C	D	E	F
1	+	−	−	+	−	+
2	+	+	−	−	+	−
3	+	+	+	−	−	+
4	−	+	+	+	−	−
5	+	−	+	+	+	−
6	−	+	−	+	+	+
7	−	−	+	−	+	+
8	−	−	−	−	−	−

Note that if an additional factor is needed, then the last column that corresponds to G is

+
+
−
+
−
−
+
−

Taguchi Designs

This appendix will provide us with the most useful two-level Taguchi designs. Even though the construction of these designs is easy, we will tabulate most of them.

L_4 Design

This design studies up to three factors with only four experimental runs.

Run	A	B	C = − AB
1	+	+	−
2	−	+	+
3	+	−	+
4	−	−	−

L_8 Design

This design studies up to seven factors with only eight experimental runs.

Run	A	B	C = − AB	D	E = − AD	F = − BD	G = ABD
1	−	−	−	−	−	−	−
2	−	−	−	+	+	+	+
3	−	+	+	−	−	+	+
4	−	+	+	+	+	−	−
5	+	−	+	−	+	−	+
6	+	−	+	+	−	+	−
7	+	+	−	−	+	+	−
8	+	+	−	+	−	−	+

L₁₂ Design

This design studies up to eleven factors with twelve experimental runs.

Run	A	B	C	D	E	F	G	H	J	K	L
1	−	−	−	−	−	−	−	−	−	−	−
2	−	−	−	−	−	+	+	+	+	+	+
3	−	−	+	+	+	−	−	−	+	+	+
4	−	+	−	+	+	−	+	+	−	−	+
5	−	+	+	−	+	+	−	+	−	+	−
6	−	+	+	+	−	+	+	−	+	−	−
7	+	−	+	+	−	−	+	+	−	+	−
8	+	−	+	−	+	+	+	−	−	−	+
9	+	−	−	+	+	+	−	+	+	−	−
10	+	+	+	−	−	−	−	+	+	−	+
11	+	+	−	+	−	+	−	−	−	+	+
12	+	+	−	−	+	−	+	−	+	+	−

L₁₆ Design

This design studies up to fifteen factors with sixteen experimental runs.

Run	A	B	C	D	E	F	G	H
1	−	−	−	−	−	−	−	−
2	−	−	−	−	−	−	−	+
3	−	−	−	+	+	+	+	−
4	−	−	−	+	+	+	+	+
5	−	+	+	−	−	+	+	−
6	−	+	+	−	−	+	+	+
7	−	+	+	+	+	−	−	−
8	−	+	+	+	+	−	−	+
9	+	−	+	−	+	−	+	−
10	+	−	+	−	+	−	+	+
11	+	−	+	+	−	+	−	−
12	+	−	+	+	−	+	−	+
13	+	+	−	−	+	+	−	−
14	+	+	−	−	+	+	−	+
15	+	+	−	+	−	−	+	−
16	+	+	−	+	−	−	+	+

Run	J	K	L	M	N	O	P
1	−	−	−	−	−	−	−
2	+	+	+	+	+	+	+
3	−	−	−	+	+	+	+
4	+	+	+	−	−	−	−
5	−	+	+	−	−	+	+
6	+	−	−	+	+	−	−
7	−	+	+	+	+	−	−
8	+	−	−	−	−	+	+
9	+	−	+	−	+	−	+
10	−	+	−	+	−	+	−
11	+	−	+	+	−	+	−
12	−	+	−	−	+	−	+
13	+	+	−	−	+	+	−
14	−	−	+	+	−	−	+
15	+	+	−	+	−	−	+
16	−	−	+	−	+	+	−

Comments on L_4, L_8, L_{12}, and L_{16} Designs

L_4 design can achieve resolution V with A, B.
L_8 design can achieve resolution V with A, B, D.
L_{16} design can achieve resolution V with A, B, D, H, P.

APPENDIX 4

Standardized Normal Distribution

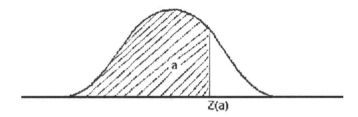

Z(a)

Cumulative Probabilities

Each four-digit entry is the area "a" under the standard normal curve from
$-\infty$ to z(a). Reference the sketch above.

z	.00	.01	.02	.03	.04	.05	.06	.07	.08	.09
.0	.5000	.5040	.5080	.5120	.5160	.5199	.5239	.5279	.5319	.5359
.1	.5398	.5438	.5478	.5517	.5557	.5596	.5636	.5675	.5714	.5753
.2	.5793	.5832	.5871	.5910	.5948	.5987	.6026	.6064	.6103	.6141
.3	.6179	.6217	.6255	.6293	.6331	.6368	.6406	.6443	.6480	.6517
.4	.6554	.6591	.6628	.6664	.6700	.6736	.6772	.6808	.6844	.6879
.5	.6915	.6950	.6985	.7019	.7054	.7088	.7123	.7157	.7190	.7224
.6	.7257	.7291	.7324	.7357	.789	.7422	.7454	.7486	.7517	.7549
.7	.7580	.7611	.7642	.7673	.7704	.7734	.7764	.7794	.7823	.7852
.8	.7881	.7910	.7939	.7967	.7995	.8023	.8051	.8078	.8106	.8133
.9	.8159	.8186	.8212	.8238	.8264	.8289	.8315	.8340	.8365	.8389
1.0	.8413	.8438	.8461	.8485	.8508	.8531	.8554	.8577	.8599	.8621
1.1	.8643	.8665	.8686	.8708	.8729	.9849	.8770	.8790	.8810	.8830
1.2	.8849	.8869	.8888	.8907	.895	.8944	.8962	.8980	.8997	.9015
1.3	.9032	.9049	.9066	.9082	.9099	.9115	.9131	.9147	.9162	.9177
1.4	.9192	.9207	.9222	.9236	.9251	.9265	.9279	.9292	.9306	.9319
1.5	.9332	.9345	.9357	.9370	.9382	.9394	.9406	.9418	.9429	.9441
1.6	.9452	.9463	.9474	.9484	.9495	.9505	.9515	.9525	.9535	.9545
1.7	.9554	.9564	.9573	.9582	.9591	.9599	.9608	.9616	.9625	.9633
1.8	.9641	.9649	.9656	.9664	.9671	.9678	.9686	.9693	.9699	.9706
1.9	.9713	.9719	.9726	.9732	.9738	.9744	.9750	.9756	.9761	.9767

z	.00	.01	.02	.03	.04	.05	.06	.07	.08	.09
2.0	.9772	.9778	.9783	.9788	.9793	.9798	.9803	.9808	.9812	.9817
2.1	.9821	.9826	.9830	.9834	.9838	.9842	.9846	.9850	.9854	.9857
2.2	.9861	.9864	.9868	.9871	.9875	.9878	.9881	.9884	.9887	.9890
2.3	.9893	.9896	.9898	.9901	.9904	.9906	.9909	.9911	.9913	.9916
2.4	.9918	.9920	.9922	.9925	.9927	.9929	.9931	.9932	.9934	.9936
2.5	.9938	.9940	.9941	.9943	.9945	.9946	.9948	.9949	.9951	.9952
2.6	.9953	.9955	.9956	.9957	.9959	.9960	.9961	.9962	.9963	.9964
2.7	.9965	.9966	.9967	.9968	.9969	.9970	.9971	.9972	.9973	.9974
2.8	.9974	.9975	.9976	.9977	.9977	.9978	.9979	.9979	.9980	.9981
2.9	.9981	.9982	.9982	.9983	.9984	.9984	.9985	.9985	.9986	.9986
3.0	.9987	.9987	.9987	.9988	.9988	.9989	.9989	.9989	.9990	.9990
3.1	.9990	.9991	.9991	.9991	.9992	.9992	.9992	.9992	.9993	.9993
3.2	.9993	.9993	.9994	.9994	.9994	.9994	.9995	.9995	.9995	.9995
3.3	.9995	.9995	.9995	.9996	.9996	.9996	.9996	.9996	.9996	.9997
3.4	.9997	.9997	.9997	.9997	.9997	.9997	.9997	.9997	.9997	.9998

Selected Percentiles

Each entry is $z(a)$ where $P[z \leq z(a)] = a$. For example, $P(z \leq 1.645) = .95$, so $z(.95) = 1.645$.

a	.10	.05	.025	.02	.01	.005	.001
z(a)	-1.282	-1.645	-1.960	-2.054	-2.326	-2.576	-3.090

a	.90	.95	.975	.98	.99	.995	.999
z(a)	1.282	1.645	1.960	2.054	2.326	2.576	3.090

Percentiles of the t Distribution

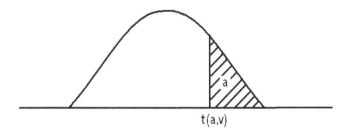

$$t(a,v)$$

Percentiles of the t distribution

df	a						
ν	.25	.10	.05	.025	.01	.005	.0005
1	1.000	3.078	6.314	12.706	31.821	63.657	636.619
2	0.816	1.886	2.920	4.303	6.965	9.925	31.599
3	0.765	1.638	2.353	3.182	4.541	5.841	12.924
4	0.741	1.533	2.132	2.776	3.747	4.604	8.610
5	0.727	1.476	2.015	2.571	3.365	4.032	6.869
6	0.718	1.440	1.943	2.447	3.143	3.707	5.959
7	0.711	1.415	1.895	2.365	2.998	3.499	5.408
8	0.706	1.397	1.860	2.306	2.896	3.355	5.041
9	0.703	1.383	1.833	2.262	2.821	3.250	4.781
10	0.700	1.372	1.812	2.228	2.764	3.169	4.587
11	0.697	1.363	1.796	2.201	2.718	3.106	4.437
12	0.695	1.356	1.782	2.179	2.681	3.055	4.318
13	0.694	1.350	1.771	2.160	2.650	3.012	4.221
14	0.692	1.345	1.761	2.145	2.624	2.977	4.140

df	a						
ν	.25	.10	.05	.025	.01	.005	.0005
15	0.691	1.341	1.753	2.131	2.602	2.947	4.073
16	0.690	1.337	1.746	2.120	2.583	2.921	4.015
17	0.689	1.333	1.740	2.110	2.567	2.898	3.965
18	0.688	1.330	1.734	2.101	2.552	2.878	3.922
19	0.688	1.328	1.729	2.093	2.539	2.861	3.883
20	0.687	1.325	1.725	2.086	2.528	2.845	3.850
21	0.686	1.323	1.721	2.080	2.518	2.831	3.819
22	0.686	1.321	1.717	2.074	2.508	2.819	3.792
23	0.685	1.319	1.714	2.069	2.500	2.807	3.768
24	0.685	1.318	1.711	2.064	2.492	2.797	3.745
25	0.684	1.316	1.708	2.060	2.485	2.787	3.725
26	0.684	1.315	1.706	2.056	2.479	2.779	3.707
27	0.684	1.314	1.703	2.052	2.473	2.771	3.690
28	0.683	1.313	1.701	2.048	2.467	2.763	3.674
29	0.683	1.311	1.699	2.045	2.462	2.756	3.659
30	0.683	1.310	1.697	2.042	2.457	2.750	3.646
40	0.681	1.303	1.684	2.021	2.423	2.704	3.551
60	0.679	1.296	1.671	2.000	2.390	2.660	3.460
120	0.677	1.289	1.658	1.980	2.358	2.617	3.373
∞	0.674	1.282	1.645	1.960	2.326	2.576	3.291

APPENDIX 6

Percentiles of the *F* Distribution

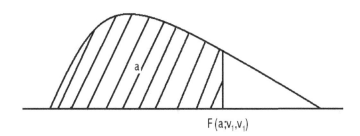

$$F(a;\nu_1,\nu_1)$$

Percentiles of the F distribution

The F distribution is skewed to the right and has two parameters:

$\nu_1 = $ numerator degrees of freedom
$\nu_2 = $ denominator degrees of freedom

The random variable F is denoted by $F(\nu_1, \nu_2)$. Entries in the tables denote the positive values $F(a;\nu_1, \nu_2)$ such that

$$P[F(\nu_1, \nu_2) \le F(a;\nu_1, \nu_2)] = a \quad \text{(reference the above sketch).}$$

The superscripted entries, like 4053^2, should be read as 405300.

Because the F distribution has two parameters, any set of tables containing even a few percentiles will be extensive. We have included the following percentiles in the pages that follow: 50th, 75th, 90th, 95th, 97.5th, 99th, and 99.9th. This set of seven percentiles is given for each of 20 different values that ν_1 and ν_2 can assume (i.e., 400 possible combinations of ν_1 and ν_2).

Percentiles less than the 50th can be found by using the following relationship:

$$F(a;\nu_1, \nu_2) = \tfrac{1}{F(1-a;\ \nu_2,\nu_1)}.$$

Please note that the order of the parameters is reversed in the two F expressions. For example, if $a = .05$, $\nu_1 = 8$, and $\nu_2 = 30$, then

$$F(.05; 8, 30) = \tfrac{1}{F(.95; 30, 8)} = \tfrac{1}{3.08} = .325.$$

229

Table A6.1. F-Distribution ($\nu_2 = 1\text{--}10$)

ν_2	a	ν_1									
		1	2	3	4	5	6	7	8	9	10
1	.500	1.00	1.50	1.71	1.82	1.89	1.94	1.98	2.00	2.03	2.04
	.750	5.83	7.50	8.20	8.58	8.82	8.98	9.10	9.19	9.26	9.32
	.900	39.9	49.5	53.6	55.8	57.2	58.2	58.9	59.4	59.9	60.2
	.950	161	200	216	225	230	234	237	239	241	242
	.975	648	800	864	900	922	937	948	957	963	969
	.990	4052	5000	5403	5625	5764	5859	5928	5982	6022	6056
	.999	4053^2	5000^2	5404^2	5625^2	5764^2	5859^2	5929^2	5982^2	6023^2	6056^2
2	.500	.667	1.00	1.13	1.21	1.25	1.28	1.30	1.32	1.33	1.34
	.750	2.57	3.00	3.15	3.23	3.28	3.31	3.34	3.35	3.37	3.38
	.900	8.53	9.00	9.16	9.24	9.29	9.33	9.35	9.37	9.38	9.39
	.950	18.50	19.00	19.20	19.20	19.30	19.30	19.40	19.40	19.40	19.40
	.975	38.50	39.00	39.20	39.20	39.30	39.30	39.40	39.40	39.40	39.40
	.990	98.50	99.00	99.20	99.20	99.30	99.30	99.40	99.40	99.40	99.40
	.999	998.0	999.0	999.0	999.0	999.0	999.0	999.0	999.0	999.0	999.0
3	.500	.585	.881	1.00	1.06	1.10	1.13	1.15	1.16	1.17	1.18
	.750	2.02	2.28	2.36	2.39	2.41	2.42	2.43	2.44	2.44	2.44
	.900	5.54	5.46	5.39	5.34	5.31	5.28	5.27	5.25	5.24	5.23
	.950	10.10	9.55	9.28	9.12	9.01	8.94	8.89	8.85	8.81	8.79
	.975	17.40	16.00	15.40	15.10	14.90	14.70	14.60	14.50	14.50	14.40
	.990	34.10	30.80	29.50	28.70	28.20	27.90	27.70	27.50	27.30	27.20
	.999	167.0	149.0	141.0	137.0	135.0	133.0	132.0	131.0	130.0	129.0
4	.500	.549	.828	.941	1.00	1.04	1.06	1.08	1.09	1.10	1.11
	.750	1.81	2.00	2.05	2.06	2.07	2.08	2.08	2.08	2.08	2.08
	.900	4.54	4.32	4.19	4.11	4.05	4.01	3.98	3.95	3.94	3.92
	.950	7.71	6.94	6.59	6.39	6.26	6.16	6.09	6.04	6.00	5.96
	.975	12.20	10.60	9.98	9.60	9.36	9.20	9.07	8.98	8.90	8.84
	.990	21.20	18.00	16.70	16.00	15.50	15.20	15.00	14.80	14.70	14.50
	.999	74.10	61.20	56.20	53.40	51.70	50.50	49.70	49.00	48.50	48.00
5	.500	.528	.799	.907	.956	1.00	1.02	1.04	1.05	1.06	1.07
	.750	1.69	1.85	1.88	1.89	1.89	1.89	1.89	1.89	1.89	1.89
	.900	4.06	3.78	3.62	3.52	3.45	3.40	3.37	3.34	3.32	3.30
	.950	6.61	5.79	5.41	5.19	5.05	4.95	4.88	4.82	4.77	4.74
	.975	10.00	8.43	7.76	7.39	7.15	6.98	6.85	6.76	6.68	6.62
	.990	16.30	13.30	12.10	11.40	11.00	10.70	10.50	10.30	10.20	10.10
	.999	47.20	37.10	33.20	31.10	29.70	28.80	28.20	27.60	27.20	26.90

Table A6.1. (cont.)

ν_2	a	1	2	3	4	5	6	7	8	9	10
6	.500	.515	.780	.886	.942	.977	1.00	1.02	1.03	1.04	1.05
	.750	1.62	1.76	1.78	1.79	1.79	1.78	1.78	1.78	1.78	1.78
	.900	3.78	3.46	3.29	3.18	3.11	3.05	3.01	2.98	2.96	2.94
	.950	5.99	5.14	4.76	4.53	4.39	4.28	4.21	4.15	4.10	4.06
	.975	8.81	7.26	6.60	6.23	5.99	5.82	5.70	5.60	5.52	5.46
	.990	13.70	10.90	9.78	9.15	8.75	8.47	8.26	8.10	7.98	7.87
	.999	35.50	27.00	23.70	21.90	20.80	20.00	19.50	19.00	18.70	18.40
7	.500	.506	.767	.871	.926	.960	.983	1.00	1.01	1.02	1.03
	.750	1.57	1.70	1.72	1.72	1.71	1.71	1.70	1.70	1.69	1.69
	.900	3.59	3.26	3.07	2.96	2.88	2.83	2.78	2.75	2.72	2.70
	.950	5.59	4.74	4.35	4.12	3.97	3.87	3.79	3.73	3.68	3.64
	.975	8.07	6.54	5.89	5.52	5.29	5.12	4.99	4.90	4.82	4.76
	.990	12.20	9.55	8.45	7.85	7.46	7.19	6.99	6.84	6.72	6.62
	.999	29.20	21.70	18.80	17.20	16.20	15.50	15.00	14.60	14.30	14.10
8	.500	.499	.757	.860	.915	.948	.971	.988	1.00	1.01	1.02
	.750	1.54	1.66	1.67	1.66	1.66	1.65	1.64	1.64	1.64	1.63
	.900	3.46	3.11	2.92	2.81	2.73	2.67	2.62	2.59	2.56	2.54
	.950	5.32	4.46	4.07	3.84	3.69	3.58	3.50	3.44	3.39	3.35
	.975	7.57	6.06	5.42	5.05	4.82	4.65	4.53	4.43	4.36	4.30
	.990	11.30	8.65	7.59	7.01	6.63	6.37	6.18	6.03	5.91	5.81
	.999	25.40	18.50	15.80	14.40	13.50	12.90	12.40	12.00	11.80	11.50
9	.500	.494	.749	.852	.906	.939	.962	.978	.990	1.00	1.01
	.750	1.51	1.62	1.63	1.63	1.62	1.61	1.60	1.60	1.59	1.59
	.900	3.36	3.01	2.81	2.69	2.61	2.55	2.51	2.47	2.44	2.42
	.950	5.12	4.26	3.86	3.63	3.48	3.37	3.29	3.23	3.18	3.14
	.975	7.21	5.71	5.08	4.72	4.48	4.32	4.20	4.10	4.03	3.96
	.990	10.60	8.02	6.99	6.42	6.06	5.80	5.61	5.47	5.35	5.26
	.999	22.90	16.40	13.90	12.60	11.70	11.10	10.70	10.40	10.10	9.89
10	.500	.490	.743	.845	.899	.932	.954	.971	.983	.992	1.00
	.750	1.49	1.60	1.60	1.59	1.59	1.58	1.57	1.56	1.56	1.55
	.900	3.28	2.92	2.73	2.61	2.52	2.46	2.41	1.38	2.35	2.32
	.950	4.96	4.10	3.71	3.48	3.33	3.22	3.14	3.07	3.02	2.98
	.975	6.94	5.46	4.83	4.47	4.24	4.07	3.95	3.85	3.78	3.72
	.990	10.00	7.56	6.55	5.99	5.64	5.39	5.20	5.06	4.94	4.85
	.999	21.00	14.90	12.60	11.30	10.50	9.92	9.52	9.20	8.96	8.75

Statistical Design of Experiments with Engineering Applications

Table A6.1. (cont.)

ν_2	a	1	2	3	4	5	6	7	8	9	10
						ν_1					
11	.500	.486	.739	.840	.893	.926	.948	.964	.977	.986	.994
	.750	1.47	1.58	1.58	1.57	1.56	1.55	1.54	1.53	1.53	1.52
	.900	3.23	2.86	2.66	2.54	2.45	2.39	2.34	2.39	2.27	2.25
	.950	4.84	3.98	3.59	3.36	3.20	3.09	3.01	2.95	2.90	2.85
	.975	6.72	5.26	4.63	4.28	4.04	3.88	3.76	3.66	3.59	3.53
	.990	9.65	7.21	6.22	5.67	5.32	5.07	4.89	4.74	4.63	4.54
	.999	19.70	13.80	11.60	10.30	9.58	9.05	8.66	8.35	8.12	7.92
12	.500	.484	.735	.835	.888	.921	.943	.959	.972	.981	.989
	.750	1.46	1.56	1.56	1.55	1.54	1.53	1.52	1.51	1.51	1.50
	.900	3.18	2.81	2.61	2.48	2.39	2.33	2.28	2.24	2.21	2.19
	.950	4.75	3.89	3.49	3.26	3.11	3.00	2.91	2.85	2.80	2.75
	.975	6.55	5.10	4.47	4.12	3.89	3.73	3.61	3.51	3.44	3.37
	.990	9.33	6.93	5.95	5.41	5.06	4.82	4.64	4.50	4.39	4.30
	.999	18.60	13.00	10.80	9.63	8.89	8.38	8.00	7.71	7.48	7.29
15	.500	.478	.726	.826	.878	.911	.933	.948	.960	.970	.977
	.750	1.43	1.52	1.52	1.51	1.49	1.48	1.47	1.46	1.46	1.45
	.900	3.07	2.70	2.49	2.36	2.27	2.21	2.16	2.12	2.09	2.06
	.950	4.54	3.68	3.29	3.06	2.90	2.79	2.71	2.64	2.59	2.54
	.975	6.20	4.76	4.15	3.80	3.58	3.41	3.29	3.20	3.12	3.06
	.990	8.68	6.36	5.42	4.89	4.56	4.32	4.14	4.00	3.89	3.80
	.999	16.60	11.30	9.34	8.25	7.57	7.09	6.74	6.47	6.26	6.08
20	.500	.472	.718	.816	.868	.900	.922	.938	.950	.959	.966
	.750	1.40	1.49	1.48	1.47	1.45	1.44	1.43	1.42	1.41	1.40
	.900	2.97	2.59	2.38	2.25	2.16	2.09	2.04	2.00	1.96	1.94
	.950	4.35	3.49	3.10	2.87	2.71	2.60	2.51	2.45	2.39	2.35
	.975	5.87	4.46	3.86	3.51	3.29	3.13	3.01	2.91	2.84	2.77
	.990	8.10	5.85	4.94	4.43	4.10	3.87	3.70	3.56	3.46	3.37
	.999	14.80	9.95	8.10	7.10	6.46	6.02	5.69	5.44	5.24	5.08
24	.500	.469	.714	.812	.863	.895	.917	.932	.944	.953	.961
	.750	1.39	1.47	1.46	1.44	1.43	1.41	1.40	1.39	1.38	1.38
	.900	2.93	2.54	2.33	2.19	2.10	2.04	1.98	1.94	1.91	1.88
	.950	4.26	3.40	3.01	2.78	2.62	2.51	2.42	2.36	2.30	2.25
	.975	5.72	4.32	3.72	3.38	3.15	2.99	2.87	2.78	2.70	2.64
	.990	7.82	5.61	4.72	4.22	3.90	3.67	3.50	3.36	3.26	3.17
	.999	14.00	9.34	7.55	6.59	5.98	5.55	5.23	4.99	4.80	4.64

Table A6.1. (cont.)

ν_2	a	1	2	3	4	5	6	7	8	9	10
						ν_1					
30	.500	.466	.709	.807	.858	.890	.912	.927	.939	.948	.955
	.750	1.38	1.45	1.44	1.42	1.41	1.39	1.38	1.37	1.36	1.35
	.900	2.88	2.49	2.28	2.14	2.05	1.98	1.93	1.88	1.85	1.82
	.950	4.17	3.32	2.92	2.69	2.53	2.42	2.33	2.27	2.21	2.16
	.975	5.57	4.18	3.59	3.25	3.08	2.87	2.75	2.65	2.57	2.51
	.990	7.56	5.39	4.51	4.02	3.70	3.47	3.30	3.17	3.07	2.98
	.999	13.30	8.77	7.05	6.12	5.53	5.12	4.82	4.58	4.39	4.24
40	.500	.463	.705	.802	.854	.885	.907	.922	.934	.943	9.50
	.750	1.36	1.44	1.42	1.40	1.39	1.37	1.36	1.35	1.34	1.33
	.900	2.84	2.44	2.23	2.09	2.00	1.93	1.87	1.83	1.79	1.76
	.950	4.08	3.23	2.84	2.61	2.45	2.34	2.25	2.18	2.12	2.08
	.975	5.42	4.05	3.46	3.13	2.90	2.74	2.62	2.53	2.45	2.39
	.990	7.08	4.98	4.13	3.65	3.34	3.12	2.95	2.82	2.72	2.63
	.999	12.00	7.76	6.60	5.70	5.13	4.73	4.44	4.21	4.02	3.87
60	.500	.461	.701	.798	.849	.880	.901	.917	.928	.937	.945
	.750	1.35	1.42	1.41	1.38	1.37	1.35	1.33	1.32	1.31	1.30
	.900	2.79	2.39	2.18	2.04	1.95	1.87	1.82	1.77	1.74	1.71
	.950	4.00	3.15	2.76	2.53	2.37	2.25	2.17	2.10	2.04	1.99
	.975	5.29	3.93	3.34	3.01	2.79	2.63	2.51	2.41	2.33	2.27
	.990	7.08	4.98	4.13	3.65	3.34	3.12	2.95	2.82	2.72	2.73
	.999	12.00	7.76	6.17	5.31	4.76	4.37	4.09	3.87	3.69	3.54
120	.500	.458	.697	.793	.844	.875	.896	.912	.923	.932	.939
	.750	1.34	1.40	1.39	1.37	1.35	1.33	1.31	1.30	1.29	1.28
	.900	2.75	2.35	2.13	1.99	1.90	1.82	1.77	1.72	1.68	1.65
	.950	3.92	3.07	2.68	2.45	2.29	2.18	2.09	2.02	1.96	1.91
	.975	5.15	3.80	3.23	2.89	2.67	2.52	2.39	2.30	2.22	2.16
	.990	6.85	4.79	3.95	3.48	3.17	2.96	2.79	2.66	2.56	2.47
	.999	11.40	7.32	5.79	4.95	4.42	4.04	3.77	3.55	3.38	3.24
∞	.500	.455	.693	.789	.839	.870	.891	.907	.918	.927	.934
	.750	1.32	1.39	1.38	1.35	1.33	1.31	1.29	1.28	1.27	1.25
	.900	2.71	2.30	2.08	1.94	1.85	1.77	1.72	1.67	1.63	1.60
	.950	3.84	3.00	2.60	2.37	2.21	2.10	2.01	1.94	1.88	1.83
	.975	5.02	3.69	3.12	2.79	2.57	2.41	2.29	2.19	2.11	2.05
	.990	6.63	4.61	3.78	3.32	3.02	2.80	2.64	2.51	2.41	2.32
	.999	10.80	6.91	5.42	4.62	4.10	3.74	3.47	3.27	3.10	2.96

Table A6.2. *F*-Distribution (ν_2 through ∞)

ν_2	a	11	12	15	20	24	30	40	60	120	∞
						ν_1					
1	.500	2.05	2.07	2.09	2.12	2.13	2.15	2.16	2.17	2.18	2.20
	.750	9.36	9.41	9.49	9.58	9.63	9.67	9.71	9.76	9.80	9.85
	.900	60.50	60.70	61.20	61.70	62.00	62.30	62.50	62.80	63.10	63.30
	.950	243	244	246	248	249	250	251	252	253	254
	.975	973	977	985	993	997	1000	1010	1010	1010	1020
	.990	6080	6110	6160	6210	6230	6260	6290	6310	6340	6370
	.999	6090^2	6110^2	6160^2	6210^2	6230^2	6260^2	6290^2	6310^2	6340^2	6370^2
2	.500	1.35	1.36	1.38	1.39	1.40	1.41	1.42	1.43	1.43	1.44
	.750	3.39	3.39	3.41	3.43	3.43	3.44	3.45	3.46	3.47	3.48
	.900	9.40	9.41	9.42	9.44	9.45	9.46	9.47	9.47	9.48	9.49
	.950	19.40	19.40	19.40	19.40	19.50	19.50	19.50	19.50	19.50	19.50
	.975	39.40	39.40	39.40	39.40	39.50	39.50	39.50	39.50	39.50	39.50
	.990	99.40	99.40	99.40	99.40	99.50	99.50	99.50	99.50	99.50	99.50
	.999	999.0	999.0	999.0	999.0	999.0	999.0	999.0	999.0	999.0	999.0
3	.500	1.19	1.20	1.21	1.23	1.23	1.24	1.25	1.25	1.26	1.27
	.750	2.45	2.45	2.46	2.46	2.46	2.47	2.47	2.47	2.47	2.47
	.900	5.22	5.22	5.20	5.18	5.18	5.17	5.16	5.15	5.14	5.13
	.950	8.76	8.74	8.70	8.66	8.63	8.62	8.59	8.57	8.55	8.53
	.975	14.40	14.30	14.30	14.20	14.10	14.10	14.00	14.00	13.90	13.90
	.990	27.10	27.10	26.90	26.70	26.60	26.50	26.40	26.30	26.20	26.10
	.999	129.0	128.0	127.0	126.0	126.0	125.0	125.0	124.0	124.0	123.0
4	.500	1.12	1.13	1.14	1.15	1.16	1.16	1.17	1.18	1.18	1.19
	.750	2.08	2.08	2.08	2.08	2.08	2.08	2.08	2.08	2.08	2.08
	.900	3.91	3.90	3.87	3.84	3.83	3.82	3.80	3.79	3.78	3.76
	.950	5.94	5.91	5.86	5.80	5.77	5.75	5.72	5.69	5.66	5.63
	.975	8.79	8.75	8.66	8.56	8.51	8.46	8.41	8.36	8.31	8.26
	.990	14.40	14.40	14.20	14.00	13.90	13.80	13.70	13.70	13.60	13.50
	.999	47.70	47.40	46.80	46.10	45.80	45.40	45.10	44.70	44.40	44.00
5	.500	1.08	1.09	1.10	1.11	1.12	1.12	1.13	1.14	1.14	1.15
	.750	1.89	1.89	1.89	1.88	1.88	1.88	1.88	1.87	1.87	1.87
	.900	3.28	3.27	3.24	3.21	3.19	3.17	3.16	3.14	3.12	3.10
	.950	4.71	4.68	4.62	4.56	4.53	4.50	4.46	4.43	4.40	4.36
	.975	6.57	6.52	6.43	6.33	6.28	6.23	6.18	6.12	6.07	6.02
	.990	9.96	9.89	9.72	9.55	9.47	9.38	9.29	9.20	9.11	9.02
	.999	26.60	26.40	25.90	25.40	25.10	24.90	24.60	24.30	24.10	23.80

Table A6.2. (cont.)

ν_2	a	11	12	15	20	24	30	40	60	120	∞
6	.500	1.05	1.06	1.07	1.08	1.09	1.10	1.10	1.11	1.12	1.12
	.750	1.77	1.77	1.76	1.76	1.75	1.75	1.75	1.74	1.74	1.74
	.900	2.29	2.90	2.87	2.84	2.82	2.80	2.78	2.76	2.74	2.72
	.950	4.03	4.00	3.94	3.87	3.84	3.81	3.77	3.74	3.70	3.67
	.975	5.41	5.37	5.27	5.17	5.12	5.07	5.01	4.96	4.90	4.85
	.990	7.79	7.72	7.56	7.40	7.31	7.23	7.14	7.06	6.97	6.88
	.999	18.20	18.00	17.60	17.10	16.90	16.70	16.40	16.20	16.00	15.70
7	.500	1.04	1.04	1.05	1.07	1.07	1.08	1.08	1.09	1.10	1.10
	.750	1.69	1.68	1.68	1.67	1.67	1.66	1.66	1.65	1.65	1.65
	.900	2.68	2.67	2.63	2.59	2.58	2.56	2.54	2.51	2.49	2.47
	.950	3.60	3.57	3.51	3.44	3.41	3.38	3.34	3.30	3.27	3..23
	.975	4.71	4.67	4.57	4.47	4.42	4.36	4.31	4.25	4.20	4.14
	.990	6.54	6.47	6.31	6.16	6.07	5.99	5.91	5.81	5.74	5.65
	.999	13.90	13.70	13.30	12.90	12.70	12.50	12.30	12.10	11.90	11.70
8	.500	1.02	1.03	1.04	1.05	1.06	1.07	1.07	1.08	1.08	1.09
	.750	1.63	1.62	1.62	1.61	1.60	1.60	1.59	1.59	1.58	1.58
	.900	2.52	2.50	2.46	2.42	2.40	2.38	2.36	2.34	2.32	2.29
	.950	3.31	3.28	3.22	3.15	3.12	3.08	3.04	3.01	2.97	2.93
	.975	4.24	4.20	4.10	4.00	3.95	3.89	3.84	3.78	3.73	3.67
	.990	5.73	5.67	5.52	5.36	5.28	5.20	5.12	5.03	4.95	4.86
	.999	11.40	11.20	10.80	10.50	10.30	10.10	9.92	9.73	9.54	9.34
9	.500	1.01	1.02	1.03	1.04	1.05	1.05	1.06	1.07	1.07	1.08
	.750	1.58	1.58	1.57	1.56	1.56	1.55	1.55	1.54	1.53	1.53
	.900	2.40	2.38	2.34	2.30	2.28	2.25	2.23	2.21	2.18	2.16
	.950	3.10	3.07	3.01	2.94	2.90	2.86	2.83	2.79	2.75	2.71
	.975	3.91	3.87	3.77	3.67	3.61	3.56	3.51	3.45	3.39	3.33
	.990	5.18	5.11	4.96	4.81	4.73	4.65	4.57	4.48	4.40	4.31
	.999	9.71	9.57	9.24	8.90	8.72	8.55	8.37	8.19	8.00	7.81
10	.500	1.01	1.01	1.02	1.03	1.04	1.05	1.05	1.06	1.06	1.07
	.750	1.55	1.54	1.53	1.52	1.52	1.51	1.51	1.50	1.49	1.48
	.900	2.30	2.28	2.24	2.20	2.18	2.16	2.13	2.11	2.08	2.06
	.950	2.94	2.91	2.85	2.77	2.74	2.70	2.66	2.62	2.58	2.54
	.975	3.66	3.62	3.52	3.42	3.37	3.31	3.26	3.20	3.14	3.08
	.990	4.77	4.71	4.56	4.41	4.33	4.25	4.17	4.08	4.00	3.91
	.999	858	8.44	8.13	7.80	7.64	7.47	7.30	7.12	6.94	6.76

Table A6.2. (cont.)

ν_2	a	ν_1 11	12	15	20	24	30	40	60	120	∞
11	.500	1.00	1.01	1.02	1.03	1.03	1.04	1.05	1.05	1.06	1.06
	.750	1.52	1.51	1.50	1.49	1.49	1.48	1.47	1.47	1.46	1.45
	.900	2.23	2.21	2.17	2.12	2.10	2.08	2.05	2.03	2.00	1.97
	.950	2.82	2.79	2.72	2.65	2.61	2.57	2.53	2.49	2.45	2.40
	.975	3.47	3.43	3.33	3.23	3.17	3.12	3.06	3.00	2.94	2.88
	.990	4.46	4.40	4.25	4.10	4.02	3.94	3.86	3.78	3.69	3.60
	.999	7.76	7.62	7.32	7.01	6.85	6.68	6.52	6.35	6.17	6.00
12	.500	.995	1.00	1.01	1.02	1.03	1.03	1.04	1.05	1.05	1.06
	.750	1.50	1.49	1.48	1.47	1.46	1.45	1.45	1.44	1.43	1.42
	.900	2.17	2.15	2.11	2.06	2.04	2.01	1.99	1.96	1.93	1.90
	.950	2.72	2.69	2.62	2.54	2.51	2.47	2.43	2.38	2.34	2.30
	.975	3.32	3.28	3.18	3.07	3.02	2.96	2.92	2.85	2.79	2.72
	.990	4.42	4.16	4.01	3.86	3.78	3.70	3.62	3.54	3.45	3.36
	.999	7.14	7.01	6.71	6.40	6.25	6.09	5.93	5.76	5.59	5.42
15	.500	.984	1.00	1.01	1.02	1.03	1.03	1.04	1.05	1.05	1.06
	.750	1.50	1.49	1.48	1.47	1.46	1.45	1.45	1.44	1.43	1.42
	.900	2.17	2.15	2.11	2.06	2.04	2.01	1.99	1.96	1.93	1.90
	.950	3.32	3.28	3.18	3.07	3.02	2.96	2.91	2.85	2.79	2.72
	.975	4.22	4.16	4.01	3.86	3.78	3.70	3.62	3.54	3.45	3.36
	.990	4.22	4.16	4.01	3.86	3.78	3.70	3.62	3.54	3.45	3.36
	.999	7.14	7.01	6.71	6.40	6.25	6.09	5.93	5.76	5.59	5.42
20	.500	.972	.977	.989	1.00	1.01	1.01	1.02	1.02	1.03	1.03
	.750	1.39	1.39	1.37	1.36	1.35	1.34	1.33	1.32	1.31	1.29
	.900	1.91	1.89	1.84	1.79	1.77	1.74	1.71	1.68	1.64	1.61
	.950	2.31	2.28	2.20	2.12	2.08	2.04	1.99	1.95	1.90	1.84
	.975	2.72	2.68	2.57	2.46	2.41	2.35	2.29	2.22	2.16	2.09
	.990	3.29	3.23	3.09	2.94	2.86	2.78	2.69	2.61	2.52.	2.42
	.999	4.94	4.82	4.56	4.29	4.15	4.01	3.86	3.70	3.54	3.38
24	.500	.967	.972	.983	.994	1.00	1.01	1.01	1.02	1.02	1.03
	.750	1.37	1.36	1.35	1.33	1.32	1.31	1.30	1.29	1.28	1.26
	.900	1.85	1.83	1.78	1.73	1.70	1.67	1.64	1.61	1.57	1.53
	.950	2,21	2.18	2.11	2.03	1.98	1.94	1.89	1.84	1.79	1.73
	.975	2.59	2.54	2.44	2.33	2.27	2.21	2.15	2.08	2.01	1.94
	.990	3.09	3.03	2.89	2.74	2.66	2.58	2.49	2.40	2.31	2.21
	.999	4.50	4.39	4.14	3.87	3.74	3.59	3.45	3.29	3.14	2.97

Table A6.2. (cont.)

ν_2	a	11	12	15	20	24	30	40	50	60	∞
30	.500	.961	.966	.978	.989	.994	1.00	1.01	1.01	1.02	1.02
	.750	1.35	1.34	1.32	1.30	1.29	1.28	1.27	1.26	1.24	1.23
	.900	1.79	1.77	1.72	1.67	1.64	1.61	1.57	1.54	1.50	1.46
	.950	2.13	2.09	2.01	1.93	1.89	1.84	1.79	1.74	1.68	1.62
	.975	2.46	2.41	2.31	2.20	2.14	2.07	2.01	1.94	1.87	1.79
	.990	2.91	2.84	2.70	2.55	2.47	2.39	2.30	2.21	2.11	2.01
	.999	4.11	4.00	3.75	3.49	3.36	3.22	3.07	2.92	2.76	2.59
40	.500	.956	.961	.972	.983	.989	.994	1.00	1.01	1.01	1.02
	.750	1.32	1.31	1.30	1.28	1.26	1.25	1.24	1.22	1.22	1.19
	.900	1.73	1.71	1.66	1.61	1.57	1.54	1.51	1.47	1.42	1.38
	.950	2.04	2.00	1.92	1.84	1.79	1.74	1.69	1.64	1.58	1.51
	.975	2.33	2.29	2.18	2.07	2.01	1.94	1.88	1.80	1.72	1.64
	.990	2.73	2.66	2.52	2.37	2.29	2.20	2.11	2.02	1.92	1.80
	.999	3.75	3.64	3.40	3.15	3.01	2.87	2.73	2.57	2.41	2.23
60	.500	.951	.956	.967	.978	.983	.989	.994	1.00	1.01	1.01
	.750	1.29	1.29	1.27	1.25	1.24	1.22	1.21	1.19	1.17	1.15
	.900	1.68	1.66	1.60	1.54	1.51	1.48	1.44	1.40	1.35	1.29
	.950	1.95	1.92	1.84	1.75	1.70	1.65	1.59	1.53	1.47	1.39
	.975	2.22	2.17	2.06	1.94	1.88	1.82	1.74	1.67	1.58	1.48
	.990	2.56	2.50	2.35	2.20	2.12	2.03	1.94	1.84	1.73	1.60
	.999	3.43	3.31	3.08	2.83	2.69	2.56	2.41	2.25	2.09	1.89
120	.500	.945	.950	.961	.972	.978	.983	.989	.994	1.00	1.01
	.750	1.27	1.26	1.24	1.22	1.21	1.19	1.18	1.16	1.13	1.10
	.900	1.62	1.60	1.55	1.48	1.45	1.41	1.37	1.32	1.26	1.19
	.950	1.87	1.83	1.75	1.66	1.61	1.55	1.50	1.43	1.35	1.25
	.975	2.10	2.05	1.95	1.82	1.76	1.69	1.61	1.53	1.43	1.31
	.990	2.40	2.34	2.19	2.03	1.95	1.86	1.76	1.66	1.53	1.38
	.999	3.12	3.02	2.78	2.53	2.40	2.26	2.11	1.95	1.76	1.54
∞	.500	.939	.945	.956	.967	.972	.978	.983	.989	.994	1.00
	.750	1.24	1.24	1.22	1.19	1.18	1.16	1.14	1.12	1.08	1.00
	.900	1.57	1.55	1.49	1.42	1.38	1.34	1.30	1.24	1.17	1.00
	.950	1.79	1.75	1.67	1.57	1.52	1.46	1.39	1.32	1.22	1.00
	.975	1.99	1.94	1.83	1.71	1.64	1.57	1.48	1.39	1.27	1.00
	.990	2.25	2.18	2.04	1.88	1.79	1.70	1.59	1.47	1.32	1.00
	.999	2.84	2.74	2.51	2.27	2.13	1.99	1.84	1.66	1.45	1.00

Some Useful Box-Behken Designs

Four-factor Box-Behnken Design

Run	A	B	C	D
1-4	± 1	± 1	0	0
5-8	0	0	± 1	± 1
9	0	0	0	0
10-13	± 1	0	0	± 1
14-17	0	± 1	± 1	0
18	0	0	0	0
19-23	± 1	0	± 1	0
24-27	0	± 1	0	± 1
28	0	0	0	0

Five-factor Box-Behnken design

Run	A	B	C	D	E
1-4	± 1	± 1	0	0	0
5-8	0	0	± 1	± 1	0
9-12	0	± 1	0	0	± 1
13-16	± 1	0	± 1	0	0
17-20	0	0	0	± 1	± 1
21-23	0	0	0	0	0
24-27	0	± 1	± 1	0	0
28-31	± 1	0	0	± 1	0
32-35	0	0	± 1	0	± 1
36-39	± 1	0	0	0	± 1
40-43	0	± 1	0	± 1	0
44-46	0	0	0	0	0

Seven-factor Box-Behnken design

Run	A	B	C	D	E	F	G
1-8	0	0	0	±1	±1	±1	0
9-16	±1	0	0	0	0	±1	±1
17-24	0	±1	0	0	±1	0	±1
25-32	±1	±1	0	±1	0	0	0
33-40	0	0	±1	±1	0	0	±1
41-48	±1	0	±1	0	±1	0	0
49-56	0	±1	±1	0	0	±1	0
57-62	0	0	0	0	0	0	0

Nine-factor Box-Behnken design

Run	A	B	C	D	E	F	G	H	J
1-8	±1	0	0	±1	0	0	±1	0	0
9-16	0	±1	0	0	±1	0	0	±1	0
17-24	0	0	±1	0	0	±1	0	0	±1
25-26	0	0	0	0	0	0	0	0	0
27-34	±1	±1	±1	0	0	0	0	0	0
35-42	0	0	0	±1	±1	±1	0	0	0
43-50	0	0	0	0	0	0	±1	±1	±1
51-52	0	0	0	0	0	0	0	0	0
53-60	±1	0	0	0	±1	0	0	0	±1
61-68	0	0	±1	±1	0	0	0	±1	0
69-76	0	±1	0	0	0	±1	±1	0	0
77-78	0	0	0	0	0	0	0	0	0
79-86	±1	0	0	0	0	±1	0	±1	0
87-94	0	±1	0	±1	0	0	0	0	±1
95-102	0	0	±1	0	±1	0	±1	0	0
103-104	0	0	0	0	0	0	0	0	0
105-112	±1	0	0	±1	0	0	±1	0	0
113-120	0	±1	0	0	±1	0	0	±1	0
121-128	0	0	±1	0	0	±1	0	0	±1
129-130	0	0	0	0	0	0	0	0	0

APPENDIX 8

Matrix Algebra

Matrix algebra plays a major role in statistical analysis. It is a necessity in multiple regression analysis. In this appendix, we will present a brief introduction to matrix algebra.

8.1 MATRICES

Definition of a Matrix: A matrix is a rectangular array of elements arranged in rows and columns. An example of a matrix is

	Column 1	Column 2
Row 1	20	12000
Row 2	60	45000
Row 3	40	40000

The elements in this particular matrix are numbers representing age (column 1) and income (column 2). Thus the element in the third row and the first column represents the age of the third person.

Square Matrix: A matrix is said to be square if the numbers of rows equals the numbers of columns. An example of a square matrix is

$$\begin{bmatrix} 2 & 3 \\ 5 & 8 \end{bmatrix}.$$

Vector: A matrix containing only one column is called a vector. An example of a vector is

$$\mathbf{V} = \begin{bmatrix} 80 \\ 10 \\ 50 \end{bmatrix}.$$

The vector $\mathbf{V}$ is a 3×1 matrix. Here, 3 denotes the number of rows, and 1 is the number of columns.

Transpose: The transpose of a matrix $\mathbf{A}$ is denoted by $\mathbf{A}'$; it is obtained by interchanging corresponding columns and rows of the matrix $\mathbf{A}$.

For example, if

$$A = \begin{bmatrix} 20 & 12000 \\ 60 & 45000 \\ 40 & 40000 \end{bmatrix},$$

then the transpose A' is

$$A' = \begin{bmatrix} 20 & 60 & 40 \\ 12000 & 45000 & 40000 \end{bmatrix}.$$

Note that the first column of A is the first row of A', and similarly the second column of A is the second row of A'.

Equality of Matrices: Two matrices A and B are said to be equal if they have the same number of columns and if all corresponding elements are equal.

For example, if

$$A = \begin{bmatrix} a_{11} & a_{12} \\ a_{21} & a_{22} \\ a_{31} & a_{32} \end{bmatrix} \quad B = \begin{bmatrix} 20 & 12000 \\ 60 & 45000 \\ 40 & 40000 \end{bmatrix},$$

then $A = B$ if and only if

$$\begin{aligned} a_{11} &= 20 & a_{12} &= 12000 \\ a_{21} &= 60 & a_{22} &= 45000 \\ a_{31} &= 40 & a_{32} &= 40000. \end{aligned}$$

Regression Examples: In the design of experiments, one basic matrix is the vector Y (output response), consisting of n design points:

$$Y = \begin{bmatrix} Y_1 \\ Y_2 \\ \vdots \\ Y_n \end{bmatrix}.$$

Another basic matrix is the X matrix, which is defined as follows for multiple regression analysis:

$$X = \begin{bmatrix} 1 & x_{11} & x_{12} & \cdots & x_{1p} \\ 1 & x_{21} & x_{22} & \cdots & x_{2p} \\ \vdots & \vdots & \vdots & & \vdots \\ 1 & x_{n1} & x_{n2} & \cdots & x_{np} \end{bmatrix},$$

where p denotes the number of effects in the regression equation.

For example, consider Example 2.3. The X and Y matrices are

$$
X = \begin{bmatrix} 1 & 1 & 1 & 1 \\ 1 & -1 & 1 & 1 \\ 1 & 1 & -1 & 1 \\ 1 & -1 & -1 & 1 \\ 1 & 1 & 1 & -1 \\ 1 & -1 & 1 & -1 \\ 1 & 1 & -1 & -1 \\ 1 & -1 & -1 & -1 \end{bmatrix} \quad Y = \begin{bmatrix} 74 \\ 70 \\ 75 \\ 68 \\ 84 \\ 81 \\ 84 \\ 80 \end{bmatrix}.
$$

8.2 MATRIX ADDITION AND SUBTRACTION

Adding or subtracting two matrices requires that they have the same number of rows and columns. The sum or difference of two matrices is another matrix whose elements each consist of the sum, or difference of the corresponding elements of the two matrices.
Suppose

$$
A = \begin{bmatrix} 20 & 40 \\ 10 & 20 \\ 30 & 50 \end{bmatrix} \quad B = \begin{bmatrix} 10 & 5 \\ 6 & 8 \\ 9 & 5 \end{bmatrix},
$$

then

$$
A + B = \begin{bmatrix} 30 & 45 \\ 16 & 28 \\ 39 & 55 \end{bmatrix}.
$$

Similarly,

$$
A - B = \begin{bmatrix} 10 & 35 \\ 4 & 12 \\ 21 & 45 \end{bmatrix}.
$$

8.3 MATRIX MULTIPLICATION

Multiplication of a Matrix by a Scalar: A scalar is a number. A matrix multiplied by a scalar is another matrix where elements are multiplied by the scalar.
For example, suppose that the matrix A is given by

$$A = \begin{bmatrix} 20 & 40 \\ 10 & 20 \\ 30 & 50 \end{bmatrix}.$$

Then

$$2A = \begin{bmatrix} 40 & 80 \\ 20 & 40 \\ 60 & 100 \end{bmatrix}.$$

Multiplication of a Matrix by a Matrix: Multiplication of a matrix by a matrix can be easily explained with an example. Consider two matrices:

$$A = \begin{bmatrix} 1 & 2 \\ 3 & 4 \end{bmatrix} \quad B = \begin{bmatrix} 2 & 5 \\ 5 & 6 \end{bmatrix}.$$

The product AB is another matrix where elements are obtained by finding the cross products of rows of A with columns of B and summing the cross products. For instance, to find the element in the first row and first column of the product AB,

$$\begin{bmatrix} 1 & 2 \\ 3 & 4 \end{bmatrix} \begin{bmatrix} 2 & 6 \\ 5 & 8 \end{bmatrix} = \begin{bmatrix} 12 & \\ & \end{bmatrix},$$

12 is determined by $(1 \times 2) + (2 \times 5)$.
To find the element of the first row and the second column of AB,

$$\begin{bmatrix} 1 & 2 \\ 3 & 4 \end{bmatrix} \begin{bmatrix} 2 & 6 \\ 5 & 8 \end{bmatrix} = \begin{bmatrix} 12 & 22 \\ & \end{bmatrix},$$

where $22 = (1 \times 6) + (2 \times 8)$.
Continuing this process, we find the product AB to be

$$AB = \begin{bmatrix} 12 & 22 \\ 26 & 50 \end{bmatrix}.$$

8.4 SPECIAL TYPES OF MATRICES

Certain special types of matrix arise in regression analysis. We shall present the most important ones.
Symmetric Matrix: If $A = A'$, A is said to be symmetric.
For example,

$$A = \begin{bmatrix} 1 & 3 \\ 3 & 2 \end{bmatrix}$$

is a symmetric matrix since $A = A'$.

Diagonal Matrix: A diagonal matrix is a square matrix whose off-diagonal elements are all zeros, such as

$$A = \begin{bmatrix} 2 & 0 & 0 \\ 0 & 5 & 0 \\ 0 & 0 & 1 \end{bmatrix}.$$

Identity Matrix: The identity matrix or unit matrix is denoted by I. It is a diagonal matrix whose elements on the main diagonal are all 1's.
For example,

$$I = \begin{bmatrix} 1 & 0 \\ 0 & 1 \end{bmatrix} \text{ or } I = \begin{bmatrix} 1 & 0 & 0 \\ 0 & 1 & 0 \\ 0 & 0 & 1 \end{bmatrix}$$

are both identity matrices.

Note that the identity matrix corresponds to the number 1 in ordinary algebra.

For any matrix A, $AI = IA$.

8.5 INVERSE OF A MATRIX

In ordinary algebra, the inverse of a number is its reciprocal. The inverse of 2 is $\frac{1}{2}$. A number multiplied by its inverse always equals 1:

$$2 \times \tfrac{1}{2} = 1.$$

In matrix algebra, the inverse of a matrix A is another matrix denoted by A^{-1}, such that

$$A^{-1}A = AA^{-1} = I,$$

where I is the identity matrix.

Examples:

1. The inverse of the matrix $A = \begin{bmatrix} 2 & 4 \\ 3 & 1 \end{bmatrix}$ is

$$A^{-1} = \begin{bmatrix} -.1 & .4 \\ .3 & -.2 \end{bmatrix}$$

since

$$A^{-1}A = \begin{bmatrix} -.1 & .4 \\ .3 & -.2 \end{bmatrix} \begin{bmatrix} 2 & 4 \\ 3 & 1 \end{bmatrix} = \begin{bmatrix} 1 & 0 \\ 0 & 1 \end{bmatrix}$$

or

$$AA^{-1} = \begin{bmatrix} 2 & 4 \\ 3 & 1 \end{bmatrix} \begin{bmatrix} -.1 & .4 \\ .3 & -.2 \end{bmatrix} = \begin{bmatrix} 1 & 0 \\ 0 & 1 \end{bmatrix}.$$

2. The inverse of a diagonal matrix

$$A = \begin{bmatrix} 2 & 0 & 0 \\ 0 & 4 & 0 \\ 0 & 0 & 1 \end{bmatrix}$$

is

$$A^{-1} = \begin{bmatrix} \frac{1}{2} & 0 & 0 \\ 0 & \frac{1}{4} & 0 \\ 0 & 0 & 1 \end{bmatrix}$$

since

$$A^{-1}A = \begin{bmatrix} \frac{1}{2} & 0 & 0 \\ 0 & \frac{1}{4} & 0 \\ 0 & 0 & 1 \end{bmatrix} \begin{bmatrix} 2 & 0 & 0 \\ 0 & 4 & 0 \\ 0 & 0 & 2 \end{bmatrix}$$

$$\equiv \begin{bmatrix} 1 & 0 & 0 \\ 0 & 1 & 0 \\ 0 & 0 & 1 \end{bmatrix}.$$

Finding the Inverse: We shall show how to obtain the inverse of small matrices.

1. If $A = \begin{bmatrix} a & b \\ c & d \end{bmatrix}$, then

$$A^{-1} = \begin{bmatrix} a & b \\ c & d \end{bmatrix}^{-1} = \begin{bmatrix} \frac{d}{\Delta} & \frac{-b}{\Delta} \\ \frac{-c}{\Delta} & \frac{a}{\Delta} \end{bmatrix},$$

where $\Delta = ad - bc$. Δ is called the determinant of the matrix A. If $\Delta = 0$, then A^{-1} does not exist.

2. If $A = \begin{bmatrix} a & 0 \\ 0 & b \end{bmatrix}$, then

$$A^{-1} = \begin{bmatrix} a & 0 \\ 0 & b \end{bmatrix}^{-1} = \begin{bmatrix} \frac{1}{a} & 0 \\ 0 & \frac{1}{b} \end{bmatrix}.$$

Index

A

B

C

D

E

F

V

Y

Z

Printed in the United States
by Baker & Taylor Publisher Services